T0324773

Complexity Theory Retrospective

Juris Hartmanis

Alan L. Selman
Editor

Complexity Theory Retrospective

In Honor of Juris Hartmanis
on the Occasion of His
Sixtieth Birthday, July 5, 1988

Springer-Verlag
New York Berlin Heidelberg
London Paris Tokyo Hong Kong

Alan L. Selman
College of Computer Science
Northeastern University
Boston, MA 02115

Printed on acid-free paper.

Photocomposed text prepared from the editor's LaTeX file.
Printed and bound by R.R. Donnelley & Sons, Harrisonburg, Virginia.
Printed in the United States of America.

9 8 7 6 5 4 3 2 1

ISBN 0-387-97350-8 Springer-Verlag New York Berlin Heidelberg
ISBN 3-540-97350-8 Springer-Verlag Berlin Heidelberg New York

Preface

In June of 1988, on the occasion of his sixtieth birthday, the participants at the Third Annual IEEE Structure in Complexity Theory Conference honored Juris Hartmanis for his many contributions to the field of complexity theory. This was most fitting, for the field takes its name from a seminal paper coauthored in 1965 by Juris Hartmanis and Richard E. Stearns titled "On the Computational Complexity of Algorithms." Hartmanis' subsequent work, reported in more than 60 papers, introduced many of the major concepts and issues of complexity theory.

The focus of this celebration was a special session of invited talks by Richard E. Stearns, Allan Borodin, and Paul Young. These speakers organized and recalled intellectual and professional trends in Hartmanis' contributions. The first three chapters of this Retrospective are the final versions of these talks, and therefore this book is at once a *festschrift* in honor of Juris Hartmanis and a culmination of the celebration that took place nearly two years ago.

There is a complementary justification for this book. All but one of the contributions to follow originated as a presentation at one of the recent meetings of the Structure in Complexity Theory Conference, and appeared in preliminary form in the conference proceedings. This conference has since its inception sought high quality technical expository presentations (in addition to the more traditional original research paper), and these expository talks have added enormously to its success. Whereas preliminary forms of research papers are customarily developed into articles that eventually appear in refereed scientific journals, no such natural forum exists for expository papers. These expositions taken together form an excellent description of much of contemporary complexity theory. Thus, I was anxious to see them published in final form in one place, and fortuitously, this proved to be the right opportunity.

The Proceedings of the Structure in Complexity Theory Conference are published by the IEEE Computer Society Press. I gratefully acknowledge the IEEE Computer Society for their support of this book.

Alan L. Selman
January, 1990

Contents

Contributors

ALLAN BORODIN
Department of Computer Science
University of Toronto
Toronto, Ontario M6C 3B7
Canada

NEIL IMMERMAN
Department of Computer and Information Science
University of Massachusetts, Amherst
Amherst, MA 01003
USA

DEBORAH JOSEPH
Computer Science Department
University of Wisconsin
Madison, WI 53706
USA

STUART A. KURTZ
Department of Computer Science
University of Chicago
Chicago, IL 60637
USA

ERIC LANDER
Whitehead Institute
9 Cambridge Center
Cambridge, MA 02142
USA

MING LI
Computer Science Department
University of Waterloo
Waterloo, Ontario N2L 3G1
Canada

STEPHEN R. MAHANEY
Computer Science Department
University of Arizona
Tucson, AZ 85721
USA

JAMES S. ROYER
School of Computer and Information Science
Syracuse University
Syracuse, NY 13210
USA

UWE SCHÖNING
Abteilung Theoretische Informatik
Universität Ulm
Oberer Eselsberg
D-7900 Ulm
Federal Republic of Germany

RICHARD E. STEARNS
Computer Science Department
State University of New York, Albany
Albany, NY 12222
USA

PAUL M.B. VITÁNYI
Centrum voor Wiskunde en Informatica
Faculteit Wiskunde en Informatica
Universiteit van Amsterdam
1098 SJ Amsterdam
The Netherlands

PAUL YOUNG
Computer Science Department
University of Washington
Seattle, WA 98195
USA

0

Introduction

Alan L. Selman[1]

I can begin no more eloquently than by quoting the master himself:

> The systematic study of computational complexity theory has
> developed into one of the central and most active research areas
> of computer science. It has grown into a rich and exciting math-
> ematical theory whose development is motivated and guided by
> computer science needs and technological advances. At the same
> time, it is clear that complexity theory deals with the quanti-
> tative laws of computation and reasoning, is concerned with
> issues and problems of direct interest to many other disciplines
> as well. It is quite likely that in the overall impact of computer
> science on our thinking, complexity theory will be recognized
> as one of the most influential intellectual contributions. [2]

As complexity theory has developed over the years, it is clear that the
field has gone through definite stages and has passed through a number
of trends and styles of research. This evolution is certainly true of Hartmanis' work as is made strikingly clear by Stearns, Borodin, and Young
in their retrospectives. Stearns describes his collaboration with Hartmanis
that resulted in the early papers. These papers focused on robustness of
the multi-tape Turing machine model, defined time and space complexity
measures, and established their basic properties. Chronologically, Borodin's
paper begins with Hartmanis' move to Cornell to Chair that university's
Computer Science Department. This was an era of intellectual development during which researchers, including Hartmanis, were concerned with
the axiomatic development of complexity measures and with the goal of
formulating correct definitions of natural complexity measures. In addition
to describing the research interests of the time, Borodin elegantly imparts
Hartmanis' role in the development of Cornell's Computer Science department and its rise to prominence.

The research fervor created by Cook's [Coo71] momentous discovery of
NP-completeness and Karp's identification of many NP-complete problems
[Kar72] crystalized the significance of nondeterminism and of completeness
in the classification of computational problems, and led to discoveries of the

---

[1]College of Computer Science, Northeastern University, Boston, MA 02115
[2]from the Introduction by Hartmanis to [Har89]

basic properties of the complexity classes (such as logarithmic and polynomial space) of the feasibly computable objects. It is a striking phenomenon even today that hundreds upon hundreds of natural computational problems fall into so few equivalence classes. This research stream, which is largely concerned with algorithms that classify specific problems, has a pronounced combinatorial flavor. Indeed, this side of complexity theory has a more pronounced combinatorial flavor than does the research stream that comprises this book.

One can discern over the past decade a growing interest in the structural properties of complexity classes. "Structural complexity theory" has a definite set of interests and has emerged as a cohesive subfield with its own rich set of problems, results, and paradigms. Structural complexity theory is characterized by its interest in global properties of complexity classes, relationships between classes, logical implications between various hypotheses about complexity classes, identification of structural properties of sets within complexity classes that affect their computational complexity, and use of relativization results as a guide to explore possible relationships.

The most outstanding illustration of the search for structure in complexity theory, and one that helped enormously to define the field, is the one that began with the observation of Berman and Hartmanis [BH77] that all known NP-complete problems are isomorphic. That is, to a polynomial time-bounded computer one NP-complete problem is the same as every other NP-complete problem. This is essentially the starting point of Young's scholarly demonstration of the long-standing consistency in Hartmanis' research themes with particular attention paid to the Isomorphism Conjecture. (This conjecture, raised by Berman and Hartmanis asserts that *all* NP-complete sets are isomorphic.) The Isomorphism Conjecture has motivated deep work by many researchers, and has sparked interest in diverse concepts such as sparse sets and polynomial degrees. Kurtz, Mahaney, and Royer, take up this theme in their article, in which they describe various endeavors motivated by this conjecture, especially their own deep work in the area.

Several aspect of structural complexity theory are taken up by the articles to follow. Immerman and Lander describe a new approach to the graph isomorphism problem that is based on Immerman's program in descriptive complexity theory. The essence of his program is to consider elements of the syntactic description of a problem as a complexity measure. Immerman has had striking success with this program. Even the proof of his well-known result that space-bounded complexity classes are closed under complements derived from his investigations in descriptive complexity theory. [3]

Li and Vitányi provide us with a compendium on Kolmogorov complexity and its interactions with structural complexity theory. The Kolmogorov

---

[3]This result was proved independently by Szelepcsényi [Sze87].

complexity of a string is the length of the shortest program that prints the string. Thus, Kolmogorov complexity measures information content. If a string is produced by a random process, then, with high probability, a program to print the string must essentially store the string itself. That most strings are random is a basic theorem of this field. This observation has led to many applications of Komogorov complexity to computational complexity theory. These are described in the article by Li and Vitányi. Hartmanis and other authors have considered the consequence of limiting the computational resource of machines that print a string, and have developed resource-bounded notions of Kolmogorov complexity that further enrich the connections between these fields.

Joseph and Young, in their jointly authored paper, study internal structural properties that individual sets frequently possess. Several of these properties when they occur in certain combinations lower complexity levels. The properties they describe have played crucial roles in many of the important results that have been obtained in recent years.

Schöning describes relationships between the complexity of determining membership and the complexity of counting. The relationships that are elementary for recursive sets seem to fail for feasibly computable sets. In addition, some of the most beautiful techniques invented in recent years involve nondeterministic procedures for counting. Various entities can be counted: number of accepting paths, number of configurations that lead to accepting configurations, number of elements of a given size of a sparse set. The well-known results of Mahaney, Hemachandra, Kadin, and Immerman (they were all were Ph.D. students of Hartmanis) all involved such nondeterministic counting, and are nicely described in Schöning's article.

These articles have not been formally refereed, but each author has graciously responded to critical comments from fellow authors. For convenience to readers, all theorem-like environments are numbered sequentially within sections. Clearly, individual authors have their own styles and idiosyncrasies. There is much to gain from digesting the individual view points to which various authors subscribe. We agreed to maintain the following consistency in notation: Standard complexity classes such as P and NP are written throughout in capital roman font. Notation for exponential time classes varies rather extensively in the research literature, so in order to set a standard, in all of the articles to follow $E = \bigcup\{\text{DTIME}(k^n) \mid k \geq 1\}$, NE $= \bigcup\{\text{NTIME}(k^n) \mid k \geq 1\}$, EXP $= \bigcup\{\text{DTIME}(2^{(p(n))} \mid p \text{ is a polynomial }\}$, and NEXP $= \bigcup\{\text{NTIME}(2^{(p(n))}) \mid p \text{ is a polynomial }\}$.

0.1 REFERENCES

[BH77] L. Berman and H. Hartmanis. On isomorphisms and density of NP and other complete sets. *SIAM J. Comput.*, 6:305–322, 1977.

[Coo71] S. Cook. The complexity of theorem-proving procedures. In *Proc.*

3rd ACM Symp. Theory of Computing, pages 151–158, 1971.

[Har89] J. Hartmanis, editor. *Computational Complexity Theory*. Volume 38 of *Procs. of Symposia in Applied Mathematics*, Amer. Math. Society, 1989.

[Kar72] R. Karp. Reducibility among combinatorial problems. In *Complexity of Computer Computations*, pages 85–104, Plenum Press, New York, 1972.

[Sze87] R. Szelepcsényi. The method of forcing for nondeterministic space. In *Bulletin of the EATCS*, pages 96–99, 1987.

1

Juris Hartmanis: The Beginnings of Computational Complexity

Richard E. Stearns[1]

I wish to commend the organizing committee for the 1988 Structures Symposium on their plan to honor Juris Hartmanis with this series of talks on his career [2]. This is a very appropriate tribute since complexity theory is now approximately 25 years old and Juris has been a prime mover in the field throughout its history. I was privileged to have worked with Hartmanis during the early period of his complexity research, and I am grateful for this opportunity to reminisce about this time period and the beginnings of complexity.

The time we spent on complexity, perhaps three and a half years, is only a short period in the history of complexity. Thus I have less ground to cover than the two talks which follow. Also, Hartmanis has already given a nice account of this time period in [Har81], and I will avoid repeating much material from that account. What I will do is discuss our results and my interpretation of their importance. The concepts we developed then are now so routinely used that their origins are sometimes forgotten.

Although the main focus of this talk is on our early complexity work, I will include some biographical information about Hartmanis from the period 1928 to 1965 and a brief discussion of our pre-complexity work. I will mainly be concerned with the time period which starts around 1960 when I first met Hartmanis and ends in 1965 when he left the General Electric Research Laboratory in Schenectady, New York to start the computer science department at Cornell. This was a very exciting time period, a time when computer science as a discipline was being invented. Indeed, computing itself was in its infancy. FORTRAN and ALGOL were recent inventions. Time sharing computing and the language BASIC were invented during that time period. The General Electric Research Laboratory where we worked had no computer until 1964. At the time we started our com-

---

[1]Computer Science Department, University at Albany - SUNY, Albany, New York 12222

[2]This paper was originally presented on June 15, 1988 at a special session of the Third Annual Symposium on Structure in Complexity Theory held in Washington, D.C.

plexity work, we had never had an opportunity to program a computer.

In my view, the central event of this time period was the development of our results on deterministic time complexity. I will discuss the importance of this event before beginning my chronological account. These complexity results were officially communicated to the world on November 11, 1964 when Hartmanis presented them in Princeton at the Fifth Annual Symposium on Switching Circuit Theory and Logical Design. This conference was an earlier incarnation of the symposium now known as FOCS (Foundations of Computer Science). The title of the paper [HS64a] was "Computational Complexity of Recursive Sequences". It was "an updated summary" of our journal paper "On the Computational Complexity of Algorithms" which was already in press [HS65]. Some differences between these versions will be pointed out later. The conference paper was given a slightly different title than the journal version so that references to the two versions could be more easily distinguished.

In 1964, the Symposium on Switching Circuit Theory and Logical Design was the one conference for computer science theory, and the conference was very well attended. However, in those days, 90 people was considered a good attendance and the conference size was not much different than today's Structure Conference, a very specialized conference. Prior to 1964, the conference had always been attached to a larger computer conference. This meeting in Princeton marked the first time the group met at a university site and made its own local arrangements. This style conference is now widely used throughout the computer science community. Hartmanis was a strong advocate of changing the conference name, and it became "Switching and Automata Theory" in 1966. In succeeding years, theory crowded out the switching theory and the conference was renamed "Foundations of Computer Science" in 1975.

Why was the paper on time complexity so important? The paper was certainly not the first paper to consider complexity issues. There were already papers within the community of logicians suggesting structural complexity [Rab59], [Rab60], [Rit63]. Perhaps our paper could be characterized as the first direct assault on the structure of deterministic time complexity. In any event, the paper uncovered the basic properties of deterministic time complexity which are so familiar today. But in addition to the results per se, the paper contained definitions and a point of view which fit very well with the more general needs of the computer science community. We will say more about the content of the paper later. Because of the visibility from the conference presentation and because the proofs could be easily understood, this paper captured the imagination of many researchers and complexity research started moving at a rapid pace. This paper also marked the first time the name "computational complexity" was used. The name quickly caught on.

Although my viewpoint is undoubtedly biased, I do not think I am exaggerating the importance of this paper. To back up this point, consider two quotations from ACM Turing Award Lectures. First I quote Steven Cook [Coo83]:

> A second early (1965) influential paper was *On the computational complexity of algorithms* by J. Hartmanis and R.E. Stearns [HS65]. This paper was widely read and gave the field its title. The important notion of complexity measure defined by the computation time on multitape Turing machines was introduced, and hierarchy theorems were proved. The paper also posed an intriguing question that is still open today. Is any irrational algebraic number (such as $\sqrt{2}$) computable in real time, that is, is there a Turing machine that prints out the decimal expansion of the number at the rate of one digit per 100 steps forever.

Next I quote Richard Karp [Kar86]:

> ...but it is the 1965 paper by Juris Hartmanis and Richard Stearns that marks the beginning of the modern era of complexity theory. Using the Turing machine as their model of an abstract computer, Hartmanis and Stearns provided a precise definition of the "complexity class" consisting of all problems solvable in a number of steps bounded by some given function of the input length n. Adapting the diagonalization technique that Turing had used to prove the undecidability of the Halting Problem, they proved many interesting results about the structure of complexity classes. All of us who read their paper could not fail to realize that we now had a satisfactory formal framework for pursuing the questions that Edmonds had raised earlier in an intuitive fashion - questions about whether, for instance, the traveling salesman problem is solvable in polynomial time.

From the viewpoint of stimulating interest in computational complexity, there is a second paper which appeared shortly after ours and also had an important influence. This was the 1967 paper of Manuel Blum [Blu67] representing work concurrent with ours. Blum presented two axioms that any reasonable complexity measure should satisfy and then described some basic properties of any such complexity measure. In particular, his results applied to time complexity as a special case. The first axiom said (intuitively) that any way of measuring complexity should assign a value to those (and only those) computations which stop. The second axiom said that it should always be decidable if a computation has a specified measured value. In the case of time complexity, a value is assigned to a halting computation by counting the steps. Thus time satisfies the first axiom. Also, one can tell if a computation takes k steps even if the computation doesn't stop. Simply run the computation for k steps and see if it stops. Thus time satisfies the second axiom.

Blum did his work as a student at MIT after attending lectures by Rabin and reading Rabin's report [Rab60]. Rabin's report was the first "early influential" paper on Cook's list. But it was Blum's paper that caught the attention of many people and gave momentum to this more abstract approach. Thus there quickly developed a two pronged attack on complexity structures.

Having given some context for our early complexity work as an event in the history of computing, I now want to discuss the work in the context of our earlier non-complexity work, and more broadly as an event in Hartmanis' career and life.

Juris Hartmanis was born on July 5, 1928 in Riga, Latvia. He lived in Latvia long enough to complete his elementary education and see most of World War II. In late 1944, the German army retreated from Latvia. Juris, together with his mother and sister, followed the German army in order to get out before the Russian army moved in. They ended up in a displaced persons camp in Hanau, Germany. While in this camp, Hartmanis completed a Latvian high school education given at the camp.

Hartmanis continued his education by entering the University of Marburg in Germany where he received a Cand.Phil. degree in Physics in 1949. In 1950, he came to the United States with his mother and sister. Under immigration laws at that time, they needed someone inside the United States to sponsor them. Sponsors were arranged for them in Kansas City, Missouri and so they came to Kansas City.

Juris applied to the University of Kansas City and was accepted. Although the degree he earned at Marburg is not the equivalent of a bachelor's degree, he had taken everything he could and had accumulated 130 credit hours. The University concluded he should enter as a graduate student. They had no physics program so Juris agreed to enter the mathematics program. He completed his M.A. degree in 1951. Although he had earned his Master's degree, his total time in college was only three and a half years.

In order to continue his education, Juris applied to several schools for the Ph.D. degree. Caltech offered him a teaching assistantship and also offered to pay his travel from Kansas City to Pasadena. They had no applied mathematics program but Juris was agreeable to doing pure mathematics. The offer of travel was the decisive factor, and Hartmanis went to Caltech. The travel cost was about $90.

While at Caltech, Juris met Ellie Rehwalt whom he would eventually marry in Schenectady. She was also born in Latvia but was raised in Germany. Hartmanis did his thesis work at Caltech under the direction of Robert P. Dilworth and received his Ph.D. in 1955. His thesis title was "Two Embedding Theorems for Finite Lattices". After Cal Tech, Juris spent two years as an instructor in mathematics at Cornell.

In the late 50's and early 60's, General Electric's Information Studies

Section under the leadership of Richard L. Shuey would bring in graduate students and young faculty for summer employment. Information Studies was part of a larger group called "Electron Physics" whose manager, James Lawson, reported directly to the director of the Research Laboratory. The atmosphere at General Electric was so free and the research so interesting that the summer employees sometimes came back as permanent employees. Among those to be recruited in this manner were Phil Lewis, Hartmanis, and myself. In 1957, Juris came to G.E. for the summer as part of that program. When the summer was over, he went to Ohio State as planned to be an assistant professor of mathematics and work with Marshall Hall. But Juris had already made up his mind to return to G.E. and Hall had made up his mind to go to Caltech, so the planned research did not materialize.

When the year at Ohio State was over, Hartmanis moved to Schenectady to become a permanent employee of the General Electric Research Laboratory. On May 18 of that same year (1958) Ellie and Juris were married in Schenectady. Included in their Schenectady years was the birth of their three children, Reneta (12/1/60), Martin (8/1/62), and Audrey (9/6/63).

I first met Hartmanis when, as a graduate student, I came to the General Electric Research Laboratory as a summer employee in 1960. When I arrived, Hartmanis had recently derived his results on partitions with the substitution property [Har61]. A quick example from [HS66a] will give the flavor of this work.

Consider the finite state sequential machine given below, with inputs {0,1}, states {1,2,3,4,5,6}, and outputs {0,1}:

	0	1	
1	4	3	0
2	6	3	0
3	5	2	0
4	2	5	1
5	1	4	0
6	3	4	0

The last column in the table describes the output of the machine for each state. This is not a finite automaton as understood today with accepting states and perhaps endmarkers. The sequential machine runs indefinitely processing input symbols and producing outputs. It was thought of as a hardware device to be implemented rather than as a language recognizer.

The motivation for this work was to represent the machine in a simple way so as to enable an efficient representation as a circuit. Analyzing the above example, Hartmanis would observe that the two partitions $\{\overline{1,2,3};\overline{4,5,6}\}$ and $\{\overline{1,6};\overline{2,5};\overline{3,4}\}$ had the "substitution property" in that the image of any partition block under any input would be wholly contained within a block of that same partition. This led to the breakup of the machine into two parallel machines as follows:

$$
\begin{array}{c|c|c}
 & 0 & 1 \\
\hline
s & t & s \\
t & s & t \\
\end{array}
$$

$$t=\{1,2,3\}, \quad s=\{4,5,6\}$$

$$
\begin{array}{c|c|c}
 & 0 & 1 \\
\hline
p & r & r \\
q & p & r \\
r & q & q \\
\end{array}
$$

$$p=\{1,6\}, \quad q=\{2,5\}, \quad r=\{3,4\}$$

Each machine is a homomorphic image of the original. Notice that knowing the states of the two machines implies a single state of the original. For example, being in state s of the first machine (representation block $\{4,5,6\}$) and state p of the second (representation block $\{1,6\}$) would imply the original machine was in $\{4,5,6\}\cap\{1,6\}$ or state 6.

Hartmanis had observed that, from a circuit design viewpoint, the simplification resulting from blocks going into blocks would be equally useful even if the original blocks were not from the same partition as the resulting blocks. This fact would be captured by the notion of a partition pair (π, τ) where blocks of π would go into blocks of τ under each input. My task for the summer of 1960 was to help push these ideas further. It was a very productive summer. When it was over, we had two operators M and m to express the important relationships, we had "information flow inequalities" to explain multiple dependencies, and we had a lattice of Mm pairs to summarize the useful partition pairs. The results made a very nice paper [SH61].

When the summer was over, I returned to Princeton where I was seeking a Ph.D. in mathematics. My thesis advisor, Harold Kuhn, introduced me to Robert Aumann who was visiting that year. Aumann steered me into working on "games without side payments", and I was able to complete my thesis work that year. Thus by the summer of 1961, I was ready for a permanent position. Because of the good experience of the previous summer, I happily accepted a position with General Electric.

Hartmanis and I continued our work on the structure of sequential machines, which continued even after starting our complexity work. We found additional applications [HS62], [HS63], and abstracted a generalization we called "pair algebra" [HS64b]. In fact, we presented a paper on pair algebra at the same conference as our complexity paper. Ultimately, we put out our work in book form [HS66a].

The work on sequential machines was elegant stuff and was frequently taught in university courses or seminars. The material was recommended as part of a course called "sequential machines" in the 1968 ACM curriculum [CUR68]. Shortly after that, interest in the subject matter declined rapidly and our book went out of print. The main use of the theory now is to distinguish the old-timers (who were educated before 1970 and remember it) from the younger people (who usually haven't heard of it). Although there may be some moral to this episode about the lasting value of research, my purpose in recalling this work is to explain the background and the culture from which we began thinking about complexity.

While working on sequential machines, we became aware of many interesting developments involving language classes and other machine models. Context-free and context-sensitive languages were the object of intense study and their basic properties were being worked out [BPS61], [Cho59], [Sch60]. Pushdown machines and Turing machines were being explored as models of computation.

Hartmanis had always been intrigued by Shannon's "information theory" [Sha48] and how information could be measured to put complexity bounds on the transmission of information. Hartmanis speculated that a similar theory could be developed to characterize computational problems and explain their computational difficulty. In 1962, the newer developments in automata theory stimulated us to start thinking about complexity. By the end of the year, our main ideas were in place. Hartmanis presents an elegant description and chronology of events of that year in [Har81].

We began our investigation on complexity by looking at regular sets and context-free sets and speculating on how they could be messed up in order to make them "harder". We derived a small body of results, most of which would have been obvious had we known any recursion theory. This was a useful exercise for us but the results were largely abandoned once we realized what the proper model should be. Although I no longer recall these pre-complexity results, they must have included some use of deterministic time because Hartmanis has recorded [Har81] our "speed-up theorem" as having been obtained in July, four months before our main insights.

By November of 1962, we had our concept of a complexity class C_T. Class C_T was to consist of those problems that could be solved in time $T(n)$ where $T(n)$ was the time required to put out the n-th bit of a sequence or (as now commonly understood) to process any input of length n. Membership in C_T is really an indication of "easyness" since C_T contains all the simple tasks and excludes the complex tasks taking more than time T. A problem is shown hard by proving it isn't in some class C_T. This corresponds to a "worst case analysis" where an algorithm is shown costly by proving that its running time sometimes (i.e. infinitely often) exceeds a specified complexity function. This is the most common approach used in the analysis of algorithms.

A second approach to complexity is to study lower bounds and demonstrate that problems are harder than lower bounds "almost everywhere". This seems more direct than the "infinitely often" approach of saying a problem isn't easy, and the existence of such problems is mathematically very interesting. However it is the "infinitely often" or "upper bound" approach that best models the methods most commonly used by the larger computer science community. The "infinitely often" approach came naturally to us since our view evolved from our interest in other resource bounded sets, namely the regular sets, context-free sets, real-time sets, and context-sensitive (or linearly bounded) sets.

As mentioned above, our first result on time complexity was our "speed-up theorem". As given in [HS65], this theorem says $C_T = C_{kT}$ for any constant k. This implies two functions T and T' define the same complexity class (i.e. $C_T = C'_T$) whenever $O(T) = O(T')$. An $O(T(n))$ algorithm proves the problem is in C_T. The result provides formal justification for analyzing algorithms only up to a constant factor. Again our model captured the common practice of the algorithms community. If you think of Hartmanis whenever you use the O-notation, then I have gotten my point across.

Although we had settled upon a very plausible complexity concept, we were lacking an understanding about the existence of complexity classes. It was in November of 1962 when we had the insight that made our complexity concept into a theory. We realized that by using multi-tape Turing machines, we could simultaneously simulate a Turing machine and impose a time limit on our simulation. Thus we could diagonalize over computations in time T using only time U whenever $inf_{n \to \infty} \frac{T(n)^2}{U(n)} = 0$. This was later improved by the Hennie-Stearns construction [HS66b] to the more familiar $inf_{n \to \infty} \frac{T(n) \cdot log(T(n))}{U(n)} = 0$. This result implied the existence of a hierarchy of complexity classes and proved that low level classes defined by familiar functions defined distinct classes. The Hennie-Stearns result was proven soon enough to be mentioned in [HS64a] but too late for [HS65].

Our breakthrough result was due in part to a deeper appreciation of the results of Hisao Yamada on real-time computation [Yam62]. Yamada showed that many commonly used functions (such as polynomials) were "real-time countable" but that other recursive functions were not. These real-time counters provided the "clocks" we needed to run our diagonalization procedures. Furthermore, Yamada was very creative with the use of multi-tape Turing machines. He provided his machines with separate tapes for each subtask, and this is what we needed to do also.

The origin of Yamada's real time idea is somewhat amusing. Yamada was enrolled in a course in Automata Theory given by Robert McNaughton at the Moore School of Electrical Engineering of the University of Pennsylvania. Yamada misunderstood a homework problem and requested a little extra time to solve it. When the assignment was turned in, McNaughton recognized that a new beast had been invented and they coined the phrase "real-time". McNaughton then became Yamada's thesis advisor. In his acknowledgement [Yam62], Yamada expresses "unreserved gratitude to Dr. R.

McNaughton...who first suggested the work reported here. During this entire period he provided guidance, suggestions, and constructive criticism...". We note that the Moore School was also the place where Myhill did his work on linear-bounded automata [Myh60].

The idea of multi-tape Turing machines seems to go back at least to the Rabin-Scott paper [RS59]. They had a finite state machine with two input tapes. There seems to have been much attention given to multiple input tapes before the implications of multiple work tapes were fully considered. For example, the conference session where Hartmanis presented our results also had a paper by Arnold Rosenberg entitled "On n-Tape Finite State Acceptors" and a paper by Seymour Ginsburg and Edwin Spanier entitled "Mappings of Languages by Two-Tape Devices".

Returning now to the main story, Hartmanis and I completed our work by showing that certain other Turing models such as multi-head or multi-dimensional tapes were easily simulated by multi-tape machines. This was to provide evidence that the complexity hierarchy does not vary much with the choice of model and thus give further justification to the multi-tape model. Prior to the publication of our results, there was considerable skepticism about Turing machines as a useful model. The skeptics reasoned that real computers do not resemble Turing machines and thus properties of Turing machines will have little bearing on real computing. Despite this skepticism, there were several papers on Turing machines at the symposium with our paper. Our complexity paper and the work which soon followed made it apparent that Turing machines were useful for studying computing despite architectural differences from real computers. The Turing model became acceptable and even fashionable for theoretical studies.

One aspect of our paper that now seems archaic was our decision to use Turing machines which generated sequences rather than Turing machines which recognized sets. The culture of that time had not yet singled out set recognition as the preferred model. There were certain technical advantages to not having to deal with input strings. However, the popularity of the recognition model was ascending rapidly. Thus in the conference version, we included a section in which the main results were restated for recognition problems.

Because of the focus on sequences, we innocently proposed a simple problem about number theory, namely "which, if any, algebraic numbers are in S_n?" We are still waiting for an answer.

When it came time to write our paper, we had to face certain presentation issues. As mathematicians, we wanted our treatment to be mathematically sound. Strictly speaking, this would require formal definitions, formal descriptions of our constructions, and formal proofs that our constructions worked. At the same time, we knew the main ideas were easy to grasp at an informal level and many were most easily explained with pictures. Furthermore, there was no established formalism for making strictly formal Turing machine proofs. We thus settled on a fairly informal style. We dutifully began [HS65] by saying that a multi-tape Turing machine was a bunch of tuples, but then went to an English description of the tuple elements. This definition was then followed by the following disclaimer:

> "For brevity and clarity, our proofs will usually appeal to the English description and will technically be only sketches of proofs. Indeed, we will not even give a formal definition of a machine operation. A formal definition of this concept can be found in [Dav58]."

The paper had seven figures. The condensed conference version had only one.

When the paper was completed, we were uncertain about which journal was most appropriate for our paper. It was unclear at that time whether automata theorists were developing new mathematics or contributing to a new discipline. More generally, people were uncertain that "computer science" was to become a respectable academic discipline independent of mathematics or electrical engineering. We sent the paper off to the *Journal of the ACM*. Two weeks later, we submitted it also to the *Transactions of the American Mathematical Society*. Hartmanis has already confessed to this "very unusual and possibly improper" action [Har81]. Today, we understand the common ground between computer science and mathematics, and would not consider such parallel submissions. But back then, we saw mathematicians and computer people as distinct groups, and we had little feeling about which group would be interested in our work.

The referee for the AMS recommended acceptance as is. The referee for the ACM made some valid criticisms of the formalism and said the paper read more like a research proposal than a finished paper. We were pleased that the paper had been accepted as proper mathematics and our choice was clear. We corrected the formalism and published in the Transactions. The AMS editor did express some concern about whether the publisher could cope with our Turing machine pictures. We assured him that we would provide drawings in a form that would satisfy the publisher and he was satisfied. (Contrary to my recollection, the account in [Har81] suggests the editor was concerned about the appropriateness of having figures.)

After the completion of our work on time complexity, Phil Lewis suggested we look at tape-bounded or space complexity. At first we told him this was too trivial. Because space complexity satisfied Blum's axioms, we knew space complexity must have the same general characteristics as time (or any other) complexity. Furthermore, the simplest tape-bound automaton was (so we thought) the linear bounded automaton since you couldn't use any less tape than that needed to write down the input. Linear bounded automata were already well understood [Myh60] and linear bounded algorithms might run in exponential time. But Phil persisted so we gave in and started investigating space.

As expected, there was a compression theorem to match the speed-up theorem. We were pleased to find that $inf_{n \to \infty} \frac{V(n)}{U(n)}$ characterized whether more sets could be computed in space U than in space V. Sometime during the investigation, we had a new idea. *We would put the input tape on a separate read-only input tape and only count worktape squares as part of the complexity.* Once we realized there were sub-linear space bounds, we eagerly plunged in. We looked at the hierarchy for four models: on-line and off-line Turing machines and on-line and off-line pushdown machines. Fretting over the nuances between on-line and off-line models is out of style today. The key model (now standard) was the off-line Turing machine. We showed that *loglogn* tape was needed before non-regular sets can be recognized and that a rich hierarchy existed beyond that point. These results were reported in [SHL65].

We continued our investigation of space complexity by working out the implications of our models for context-free and context-sensitive languages [LSH65]. We looked at all the models from [SHL65] but mention here only some results for the off-line Turing machine case. We found a rich hierarchy of context-free languages between *loglogn* space and *logn* space. We found that $(logn)^2$ space was sufficient for all context-free languages. We left open the question as to whether more than *logn* space was needed.

Possibly the most influential idea in our paper [LSH65] was our proof that recognizing context-free languages on a Turing machine requires at most $(logn)^2$ space. Our algorithm employed a divide and conquer technique whereby a substring of roughly half size would be parsed and then the substring would be replaced by a nonterminal to give a second half-sized string to be analyzed. This divide and conquer theme was picked up later by Savitch in the famous paper [Sav70] showing $L(n)$ nondeterministic space could be simulated in $L(n)^2$ deterministic space. In the proof of his main theorem, Savitch says [Sav70], p. 179: "The algorithm is similar to that used by Lewis, Stearns and Hartmanis [LSH65] to show that every context-free language is accepted by a deterministic Turing machine with storage $(logn)^2$."

We were invited to give a paper [HLS65] at the 1965 IFIP Conference in New York and we presented a summary of results from [SHL65] and [LSH65]. This conference marked the end of our collaboration since it was then that Hartmanis announced his plan to go to Cornell to start a computer science department.

The results from this early time period are now such common knowledge that their origins are sometimes forgotten. Yet simply restating them does not capture their impact. By means of this informal account, I have tried to create an understanding of the culture of the time and to explain the impact within the context of that culture. I also hope I have conveyed the importance of finding the right models. Finding a good model is often a significant intellectual achievement even though, afterward, the model can be expressed by a short definition. I consider myself very privileged to have participated in these early discoveries and to have had an association with

16   Richard E. Stearns

Juris Hartmanis. From these beginnings, Hartmanis became the driving force in the development of complexity theory. The diversity of results presented at this conference is ample evidence of his success.

1.1 REFERENCES

[Blu67]   M. Blum. A machine-independent theory of the complexity of recursive functions. *J. ACM*, 14,2:322–336, April 1967.

[BPS61]   Y. Bar-Hillel, M. Perles, and E. Shamir. On formal properties of simple phrase structure grammars. *Z. Phonetik, Sprach. Komm.*, 14:143–179, 1961.

[Cho59]   N. Chomsky. On certain formal properties of grammars. *Information and Control*, 2:136–167, 1959.

[Coo83]   S.A. Cook. An overview of computational complexity. *Communications of ACM*, 26,6:401–402, June 1983.

[CUR68]   *Curriculum 68 – Recommendations for Academic Problems in Computer Science – A Report of the ACM Curriculum Committee on Computer Science*, 1968. Communications of ACM, vol. 11,3, page 151-197.

[Dav58]   M. Davis. *Computability and Unsolvability*. McGraw-Hill, New York, 1958.

[Har61]   J. Hartmanis. On the state assignment problem for sequential machines I. *IRE Trans. Elec. Comp.*, 10:157–165, 1961.

[Har81]   J. Hartmanis. Observations about the development of theoretical computer science. *Annals of the History of Computing*, 3,1:42–51, January 1981.

[HLS65]   J. Hartmanis, P.M. Lewis, and R.E. Stearns. Classification of computations by time and memory requirements. In *Proc. of IFIP Congress*, pages 31–35, 1965.

[HS62]   J. Hartmanis and R.E. Stearns. Some dangers in state reduction of sequential machines. *Information and Control*, 5,3:252–260, September 1962.

[HS63]   J. Hartmanis and R.E. Stearns. A study of feedback and errors in sequential machines. *IEEE Transactions on Electronic Computers*, EC-12,3, June 1963.

[HS64a]   J. Hartmanis and R.E. Stearns. Compuational complexity of recursive sequences. In *Proc. Fifth Annual Symp. on Switching Theory and Logical Design*, Princeton, NJ, 1964.

[HS64b]   J. Hartmanis and R.E. Stearns. Pair algebra and its application to automata theory. *Information and Control*, 7,4:485–507, December 1964.

[HS65]   J. Hartmanis and R.E. Stearns. On the computational complexity of algorithms. *Transactions of American Mathematical Society*, 117,5:285–306, May 1965.

[HS66a]   J. Hartmanis and R.E. Stearns. *Algebraic Structure Theory of Sequential Machines*. Prentice-Hall, 1966.

[HS66b]   F.C. Hennie and R.E. Stearns. Two-tape simulation of multitape turing machines. *Journal of ACM*, 13,4:533–546, October 1966.

[Kar86]   R.M. Karp. Combinatorics, complexity, and randomness. *Communications of ACM*, 29,2:103, February 1986.

[LSH65]   P.M. Lewis, R.E. Stearns, and J. Hartmanis. Memory bounds for recognition of context free and context sensitive languages. In *Proc. 6th Annual Symp. on Switching Circuit Theory and Logical Design*, pages 191–202, October 1965.

[Myh60]   J. Myhill. *Linear Bounded Automata*. Technical Report No. 60-22, University of Pennsylvania, 1960.

[Rab59]   M.O. Rabin. Speed of computation and classification of recursive sets. In *Third Convention Science Society*, pages 1–2, Israel, 1959.

[Rab60]   M.O. Rabin. *Degree of Difficulty of Computing a Function and a Partial Ordering of Recursive Sets*. Technical Report No. 2, Hebrew University, Jerusalem, 1960.

[Rit63]   R.W. Ritchie. Classes of predictably computable functions. *Trans. American Mathematical Society*, 106:139–173, 1963.

[RS59]   M.O. Rabin and D. Scott. Finite automata and their decision problems. *IBM Journal on Research*, 3:115–125, 1959.

[Sav70]   W.J. Savitch. Relations between nondeterministic and deterministic tape complexities. *J. Comp. and Sys. Sciences*, 4:177–192, 1970.

[Sch60]   S. Scheinberg. Note on the boolean properties of context-free languages. *Information and Control*, 3:373–375, 1960.

[SH61]   R.E. Stearns and J. Hartmanis. On the state assignment problem for sequential machines II. *IRE Transactions on Electronic Computers*, EC-10,4:593–603, December 1961.

[Sha48]   C.E. Shannon. Mathematical theory of communication. *Bell Sys. Tech. Journal*, 17:379–423,623–658, 1948.

[SHL65]   R.E. Stearns, J. Hartmanis, and P.M. Lewis. Hierarchies of memory limited computations. *Proc. 6th Annual Symp. on Switching Circuit Theory and Logical Design*, 179–190, October 1965.

[Yam62]   H. Yamada. Real-time computation and recursive functions not real-time computable. *IRE Trans. Elec. Comp.*, EC-11:753–760, 1962.

2

Juris Hartmanis: Building a Department—Building a Discipline

Allan Borodin[1]

In January, 1988, Alan Selman called and asked me to talk at the 3rd Annual Structures Conference about "Juris Hartmanis: The Middle Years". Somehow I forgot that a talk in June implies a written paper in March—hence the very preliminary nature of that paper. It is almost two years later and I feel that I am still working on a preliminary version. I also forgot just how much my views about research and about computer science as a discipline were directly inherited from my supervisor. I interpreted the "middle years" to be the years from 1965 (when Juris came to Cornell as Chairman of a newly formed Computer Science Department) until 1978 (when the SIAM monograph [JHa78] appears). Given Juris' energy and enthusiasm for research, I fully expect that in another twenty years we will be doing this again, by which time the "middle years" will have become part of the "early years". So I chose a title which I think better suggests the profound impact Juris Hartmanis has had on our discipline. Beyond his seminal and ongoing contributions to the field of complexity, Hartmanis was able to use his research reputation not only to develop a department but, moreover, to strongly influence the direction of computer science as a distinct discipline.

The idea of a Computer Science Department at Cornell was primarily the inspiration of three Cornell faculty members: Richard Conway of Industrial Engineering and Operations Research, Anil Nerode of Mathematics and Robert Walker of Mathematics. Conway was somehow able to arrange a $1,000,000 grant from the Sloan Foundation pending the appointment of a chair. Nerode had established a computer science seminar series in the Center for Applied Mathematics so that prospective candidates would at least have an audience and maybe even the illusion of intense activity. The Conway-Nerode-Walker triumvirate was able to arrange offers to Hartmanis, Pat Fischer and Gerry Salton, who collectively decided to accept.

Hartmanis became the Cornell Chairman having had a little teaching

---

[1]Department of Computer Science, University of Toronto

experience at Cornell and Ohio State but no administrative experience. Conway-Nerode-Walker were on hand to offer advice but only if asked. Hartmanis was the Chairman of (what was then) a very questionable and undefined discipline. Outspoken physicists couldn't understand why respected mathematicians were supporting a Ph.D. program for the study of Fortran. Moreover, the chairman of the unknown department would have to report to two deans, Arts and Science and Engineering. Here, Hartmanis used his training in Boolean algebra; rather than seek approval from $D_1 \wedge$ from D_2 for a particular request, it sufficed to seek approval from $D_1 \vee$ from D_2 depending on who was most likely to approve. Nothing succeeds like success. John Hopcroft arrived in September 1967, Bob Constable in 1968 and David Gries in 1969. They are all still at Cornell. Other faculty members recruited by Hartmanis in his first term as chairman (1965–1971) include Ken Brown, John Dennis, Ellis Horowitz, George Moré, Alan Shaw, Bob Tarjan and Peter Wegner.

Hartmanis ran the department in what might be called a democratic kingdom. There were no organized faculty meetings; rather, after lunch Juris informally held court at the Statler Faculty Club. There everyone got to express opinions on the topic of the day until a consensus was formed. There was a definite correlation between this consensus and Juris' position on the topic. The style was remarkably informal but also remarkably efficient.

I came to Cornell in the fall of 1966 as part of the first Computer Science graduate class. Of course, as a student, I didn't get to observe Hartmanis hold court in Statler or surgically operate on the administration, but I can extrapolate from his performance as a teacher. The excitement of the new discoveries concerning complexity classes (described by Dick Stearns [RSt] and by Hartmanis [JHa81]) was extended to his teaching. We learned much more than results. We learned that results take place in a context; we learned about the importance of appropriate models of computation, the importance of asking the right question, the importance of intuitive as well as more formal proofs, the importance of a surprising or elegant result, the importance of a good conjecture, the importance of having an opinion about what is and what is not important within that context. Mostly we learned that research was exciting and that we could do it.

In the same way that Hartmanis had strong convictions about the direction of complexity theory, we can observe his determination in setting directions for computer science as a whole. It would have to be based on strong foundations, it would have to gain the respect of other disciplines (especially mathematics), and one should expect (and welcome) many surprising developments. Cornell would initially stay focused on a few areas and emphasize computer science as a science.

It is not a coincidence that this view was adopted by many other universities. Hartmanis was officially on the review and advisory committee for twelve universities (and probably unofficially for another dozen), he served on at least seven editorial boards as well as numerous national committees. (And he continues to do so.) Probably his most effective strategy was to plant his students throughout the academic world. (Juris' PHD students currently hold faculty positions at Aarhus, Chicago, Cornell, Dartmouth, Florida State, Kentucky, Illinois, Northeastern, Princeton, Rochester, Toronto, University of Massachusetts and Waterloo.)

From a more technical perspective, we can clearly see the complexity directions that Hartmanis was advocating. His monograph "Feasible Computations and Provable Complexity Properties" [JHa78] is a beautiful exposition of this direction. Indeed, to me it defines "Structures". Since Paul Young [PYo] discusses many of these results in relative detail, including the elegant conjectures and results concerning isomorphism and sparsity, I will only briefly mention some other results so as to give a sense of the activity during this time.

It seems appropriate for me to begin where Richard Stearns [RSt] ends his discussion of The Beginnings of Computational Complexity, with the study of space bounded complexity classes and context free language recognition. In this regard, I recall the two papers ([JH68] and [JH69]) written with Herb Shank. The first paper shows that for any infinite language $L \subseteq$ {primes}, L is not a CFL. While this result (also proven independently by Schutzenberger) is part of formal language theory, the second paper belongs to what is today called computational number theory. In a field where lower bounds are virtually absent, Hartmanis and Shank were able to establish two lower bounds for the space complexity of recognizing primes, namely that recognition of {primes} requires $\Omega(\log n)$ space using the usual two-way read only input model, and requires $\Omega(n)$ space if one restricts the input to a one-way read only tape. As is often the case in his papers, we also find fundamental (and still unresolved) conjectures. Here, Hartmanis and Shank conjecture that recognition of {primes} requires $\Omega(n)$ space but that there exists an infinite $L \subseteq$ {primes} which can be recognized in less than linear space. Only recently have I seen any progress on these conjectures, notably the Adelman and Kompella [LK88] results on depth-size complexity of randomized unbounded fan-in circuits for number theoretic problems (including primality) which can be translated to yield a "non uniform" $O(\sqrt{n} \text{ poly} \log n)$ space upper bound. In contrast to the Hartmanis and Shank conjecture, they also conjecture circuit results which would yield uniform sublinear space bounds. We see, in retrospect, that recent interest in the parallel complexity of number theoretic problems has some roots in these papers.

The relation between space and parallel time complexity (to which I have just alluded) was itself a subject in which Hartmanis made an important contribution. Following Pratt and Stockmeyer's [VL76] PSPACE characterization of the power of vector machines, Hartmanis and his student Janos Simon [JJ74] gave an elegant alternative characterization of PSPACE as PTIME(+, -, ×, Bool); that is those sets recognizable (or functions computable) by a polynomial time bounded RAM whose atomic operations are integer addition, subtraction and multiplication as well as the (bitwise) Boolean operations. Here it is the combination of the boolean operations and multiplication that provides the opportunity for large scale (exponential) parallelism. This contrasts nicely with the characterization of P (= sequential polynomial time) as PTIME[+, -, bool]; that is, the same RAM model without multiplication as a primitive. They also show that PTIME[+, -, ×, bool] = NPTIME[+, -, ×, bool]; that is, the determin-istic and nondeterministic classes are equivalent whereas Cook's [SCo71] celebrated P vs NP question can be formulated as PTIME[+, -, bool] vs NPTIME[+, -, bool]. Even in an area which today might be considered "parallel complexity theory", we see the structural emphasis that Hart-manis gives his results. The title and much of the emphasis of the paper is on the relative power of models. Motivated by Pratt & Stockmeyer and Hartmanis & Simon, the relation between space and parallel time is further developed in a number of papers (see, for example, Borodin [ABo77] for the circuit model and Goldschlager [LGo82] for the SIMDAG = CRCW-PRAM model and the "parallel computation thesis"). The results of Simon [JSi79] and Bertoni, et al [AGN81] show that the Hartmanis and Simon RAM model can be extended to allow integer division.

As a final comment of the early development of a parallel complexity theory, it is worth noting the identification and study of tape reversals as an interesting complexity measure by Fischer, Hartmanis and Blum [PJM68] and Hartmanis [JHa68]. While tape reversals may not have had the natural appeal of the time or space measures, we later see in Pippenger [NPi79] the importance and relevance of this measure in defining the class NC, problems solvable in polylog parallel time using a polynomial number of processors.

Another aspect of the research activity during this period relates to Hart-manis' desire to maintain perspective within the field. To this end, he and John Hopcroft collaborated on two papers which provided a much needed unification; a clear exposition of undecidedability results in formal language theory [JJ70] and the basic results (both machine dependent and axiomatic machine independent) of complexity theory [JJ71]. As I remember that pe-riod, Hartmanis alternated between trying to unify the field with Hopcroft and trying to dismember him (and anyone else) on the other side of the volleyball net.

Returning to "Feasible Computations and Provable Complexity Proper-ties", we see not only a unification of ideas but moreover a spirit of research that defines Structures. A wealth of results are presented both in terms of formal theorems and in terms of informal explanations which motivate and capture the goals and importance of the Structures area. A central re-search theme during this period, one very much part of "Structures", was the importance of representation (in particular, succinctness results) and of provability.

The focus of Chapter 6 [JHa78] is Hartmanis' long standing interest in a constructive complexity theory, analogous to Fischer's [PFi65] constructive recursive function theory. If one adopts the view of complexity classes based on provable inclusion in the class (i.e. there is a machine for the language whose complexity is provably within the required bound) then the structure of this constructive complexity theory looks quite different from the classical theory. For example, we find that for every non decreasing g, there exist arbitrarily large bounds t such that the class of languages in the provable class induced by the bound $g \circ t$ is properly contained in the classical class induced by the bound t. It follows, to quote Hartmanis, that the "originally surprising gap theorem ... is a result about a class of languages whose defining properties cannot be verified formally". (It was too late, I had already graduated.) Perhaps the most interesting result in this regard is Young's [PYo77] program optimization among provably equivalent programs, in contrast to Blum's [MBl67a] speedup theorem. Such results hold for all proof systems and, generally speaking, for all complexity measures satisfying the Blum [MBl67a] axioms. There are also a few results specific to tape and time bounded classes, and it is still an open conjecture as to whether constructible time bounds induce the same classes, in the provable and classical sense of classes.

Related to this is the issue of the independence of complexity structure from given formalized logical theories. Chapter 7 of [JHa78] provides a number of examples of this type of result. Not far removed are the relativization results pioneered by Baker, Gill and Solovay [TJR75], Baker being a Hartmanis student.

And finally we mention the issue of succinctness, another important aspect of the general issue of representations and models that (as we have seen) underlines so much of Hartmanis' research activity during this "middle period". Relative succinctness of formal proof systems originates in the work of Gödel [KGo36]. In Chapter 5 of [JHa78], we see a very intuitive and elegant computational development showing that "In every formalization, infinite sets of trivial theorems will require very long proofs". Perhaps of more direct interest to computer scientists are relative succinctness results concerning models of computation [MBl67b] and representations of formal languages [AM71], [ET77]. For example, Schmidt (another Hartmanis' student) and Szymanski show that for unambiguous context free languages, there is no recursive function which bounds the size of the minimal unambiguous context free grammar representation in terms of the size of the minimal representation using ambiguous grammars. In [JHa80], we see a unified development of these language succinctness results and, moreover, we see a strong connection to the work concerning provable complexity classes. Hartmanis establishes results concerning the relative succinctness of finite languages represented by finite automata relative to representation by Turing machines. If the Turing machine acceptors come with proofs that give explicit bounds on the cardinality of the finite set being accepted, then there is a recursive bound on the relative succinctness. If the Turing machines only come with proofs that the set accepted is finite (or with no proof) then there is no recursive bound on the relative succinctness. More results relating succinctness and provability can be found in Baker and Hartmanis [TJ79].

Having mentioned myself, Baker, Schmidt and Simon, this may be the appropriate place to mention (in chronological order) the thesis work of the other Hartmanis students of this period. (Thus far, Hartmanis has had sixteen Ph.D. graduates of which eight were in the 1969–1978 period). Forbes Lewis studied the structure of formal language decision problems; Ed Reingold studied computation trees with different primitives (e.g. linear comparisons); Len Berman studied the isomorphism conjecture; Dexter Kozen studied the complexity of finitely presented algebras. There is a great diversity in these topics but there are also common themes. In addition to many quotable results, there are an abundance of good (and still open) conjectures. Erik Schmidt recalls Juris' advice on what constitutes a good result around which one could build a thesis: "It doesn't have to be revolutionary and it doesn't have to be very difficult, but *it must tell a story*".

Lane Hemachandra made the following comments concerning the "very preliminary version" of this paper. "One point that struck me in the paper was your mention of the fact that Professor Hartmanis has had 16 advisees. I found this totally surprising. Sixteen is a small finite constant—though I had never really thought much about how many students he's had, based on the fact that wherever one turns (in structural complexity theory) one bumps into the work of Hartmanis students, I would have guessed that there were far more. This reflects, I think, the fact that Professor Hartmanis has a really unparalleled talent for teaching students to step back, look at things in broad perspective, and to do the technical details that are a part of research only when you know the meaning and goal of what you're attempting."

Obviously, in such a short review I haven't done justice to Hartmanis' technical contributions during this time period. Nor have I done justice to all the people whose work significantly influenced or was influenced by Hartmanis. As you all know, Juris is a forceful and persuasive expositor, not only for his own results and for his students but for the field in general. His expositions need no further elaboration from me. Moreover, the technical contributions are only part of the story I wanted to tell.

The other part of the story brings us back to the title. I commented earlier that Juris was officially on the review or advisory committees for twelve universities. Twelve is also a small finite constant which doesn't tell the story. Juris certainly knew and could clearly articulate the meaning and goals of what he was attempting to develop in terms of the Cornell department, and he was and continues to be very persuasive about the meaning and goals of Computer Science. Having established strong foundations at Cornell, Juris then became a forceful advocate for developing the systems component of the discipline. It is a pattern of development that can be observed in so many departments throughout North America and Europe. Computer Science is now an accepted discipline. We even appear to be weathering a period of declining undergraduate enrollment with almost as much success as we "enjoyed" in the days of seemingly unbounded enrollment. (Administrators reason that students have finally discovered that computer science is a demanding field of study.) We now have joint instructional and research programs with physicists, as well as with mathematicians, electrical and mechanical engineers, cognitive scientists and linguists, etc. In short, we have the kind of discipline that Juris Hartmanis had envisioned, articulated and successfully advocated.

Acknowledgements: Many thanks to Bob Constable, Richard Conway, David Gries, John Hopcroft and Anil Nerode for sharing their recollections. Thanks also to Erik Schmidt and Lane Hemachandra for their comments. And finally, I thank Alan Selman and the Program Committee for this opportunity to express my appreciation to Juris for my academic roots.

2.1 REFERENCES

[ABo77] A. Borodin. On relating time and space to size and depth. *SIAM J. COMPUT.*, 6(4):733–744, December 1977.

[AGN81] A. Bertoni, G. Mauri, and N. Sabadini. A characterization of the class of functions computable in polynomial time on random access machines. In *Proc. 13th Annual ACM Symposium on Theory of computing*, pages 168–176, May 1981.

[AM71] A. Meyer and M. Fischer. Economy of description by automata, grammars and formal systems. In *Conference Record, IEEE 12th Annual Symposium on Switching and Automata Theory*, pages 188–190, 1971.

[ET77] E. Schmidt and T. Szymanski. Succinctness of descriptions of unambiguous context-free languages. *SIAM J. Comput.*, 6:547–553, 1977.

[JH68] J. Hartmanis and H. Shank. On the recognition of primes by automata. *JACM*, 15(3):382–389, July 1968.

[JH69] J. Hartmanis and H. Shank. Two memory bounds for the recognition of primes by automata. *Mathematical Systems Theory*, 3(2):125–129, 1969.

[JHa68] J. Hartmanis. Tape reversal bounded turing machine computations. *JCSS*, 2(2):117–135, August 1968.

[JHa78]   J. Hartmanis. Feasible computations and provable complexity properties. 1978. *CBMS—NSF Regional Conf. Series in Applied Mathematics 30*, SIAM Monograph, 62 pages.

[JHa80]   J. Hartmanis. On the succinctness of different representations of languages. *SIAM. J. Comput.*, 9(1):114–120, February 1980.

[JHa81]   J. Hartmanis. Observations about the development of theoretical computer science. *Annals of the History of Computing*, 3(1):42–51, 1981.

[JJ70]   J. Hartmanis and J. Hopcroft. What makes some language theory problems undecidable. *JCSS*, 4(4):368–376, August 1970.

[JJ71]   J. Hartmanis and J. Hopcroft. An overview of the theory of computational complexity. *J. Assoc. Comput. Mach.*, 18:444–475, 1971.

[JJ74]   J. Hartmanis and J. Simon. On the power of multiplication on random access machines. In *Proc. IEEE 15th Symp. on Switching and Automata Theory*, pages 13–23, October 1974.

[JSi79]   J. Simon. Division is good. In *20th Annual Symposium on Foundations of Computer Science*, pages 411–420, October 1979.

[KGo36]   K. Godel. Uber die lange von beweisen. *Ergebnisse eines mathematischen Kolloquiuns*, 7:23–24, 1936.

[LGo82]   L. Goldschlager. Synchronous parallel computation. *Assoc. Comput. Mach.*, 29(4):1073–1086, 1982.

[LK88]   L. Adleman and K. Kompella. Using smoothness to achieve parallelism (abstract). In *Proc. 20th Annual ACM Symposium on Theory of Computing*, pages 528–538, May 1988.

[MBl67a] M. Blum. A machine-independent theory of the complexity of recursive functions. *J. Assoc. Comput. Mach.*, 14:322–336, 1967.

[MBl67b] M. Blum. On the size of machines. *Information and Control*, 11:257–265, 1967.

[NPi79]   N. Pippenger. On simultaneous resource bounds (preliminary version). In *Proc. 20th IEEE Found. of Comput. Sci.*, pages 307–311, 1979.

[PFi65]   P. Fischer. Theory of provable recursive functions. *Trans. Amer. Math. Soc.*, 117:494–520, 1965.

[PJM68] P. Fischer, J. Hartmanis, and M. Blum. Tape reversal complexity hierarchies. In *IEEE Conference Record of the 1968 Ninth Annual Symposium on Switching and Automata Theory*, pages 373–382, October 1968.

[PYo]   P. Young. Juris Hartmanis: fundamental contributions to isomorphism problems. In A. Selman, editor *Complexity Theory Retrospective*, pages 28–58, Springer-Verlag, 1990.

[PYo77] P. Young. Optimization among provably equivalent programs. *J. Assoc. Comput. Math.*, 24:693–700, 1977.

[RSt]   R. Stearns. Juris Hartmanis: the beginnings of computational complexity. In A. Selman, editor *Complexity Theory Retrospective*, pages 5–18, Springer-Verlag, 1990.

[SCo71] S. Cook. The complexity of theorem proving procedures. In *Proc. 3rd Annual ACM Symposium on Theory of Computing*, pages 151–158, May 1971.

[TJ79]   T. Baker and J. Hartmanis. Succinctness, verifiability and determinism in representations of polynomial-time languages. In *Proc. 20th IEEE Found. of Comput. Sci.*, pages 392–396, 1979.

[TJR75] T. Baker, J. Gill, and R. Solovay. Relativizations on the P =?NP question. *SIAM J. Comput.*, 431–442, 1975.

[VL76]   V. Pratt and L. Stockmeyer. A characterization of the power of vector machines. *J. Comput. System Sci.*, 12:198–221, 1976.

3

Juris Hartmanis: Fundamental Contributions to Isomorphism Problems

Paul Young[1]

ABSTRACT In this paper we survey Juris Hartmanis' contributions to isomorphism problems. These problems are primarily of two forms. First, isomorphism problems for restricted programming systems, including the *Hartmanis-Baker conjecture* that all polynomial time programming systems are polynomially isomorphic. Second, the research on isomorphisms, and particularly polynomial time isomorphisms for complete problems for various natural complexity classes, including the *Berman-Hartmanis conjecture* that all sets complete for NP under Karp reductions are polynomially isomorphic. We discuss not only the work of Hartmanis and his students on these isomorphism problems, but we also include a (necessarily partial and incomplete) discussion of the the impact which this research has had on other topics and other researchers in structural complexity theory.

3.1   Introduction

Juris Hartmanis' research, and almost all research in structural complexity theory on isomorphism problems, has centered around two conjectures made in the 1970's by Hartmanis and his students, Ted Baker and Leonard Berman. The first of these conjectures, *the Hartmanis-Baker conjecture*, is that all polynomial time programming systems are polynomially isomorphic. The second of these conjectures, *the Berman-Hartmanis conjecture*, is that all sets which are Karp complete for NP are polynomially isomorphic.

Because these conjectures were preceded and partially motivated by similar theorems in recursion theory, it will be useful to begin our discussion of Hartmanis' work with a brief review of the recursion and complexity theoretic setting which formed the background for his research on these problems.

---

[1]Computer Science Department, University of Washington, Seattle, WA 98195

In his fundamental paper published in 1958, ([Ro-58]), Hartley Rogers defined an enumeration

$$\phi_0, \phi_1, \phi_2, \ldots$$

of the partial recursive functions to be an *acceptable indexing* or a *Gödel numbering* of the partial recursive functions if it has a computable universal function, $UNIV$, that is, a partial function of two arguments such that for all i and x, $UNIV(i, x) = \phi_i(x)$, and also computable (total) s-m-n functions S_m^n such that for all i, for all m-tuples $\overline{x}$ and all n-tuples $\overline{y}$, $\phi_{S_m^n(i,\overline{x})}(\overline{y}) = \phi_i(\overline{x}, \overline{y})$. Because they model so many features of programming systems and models of computation, Gödel numberings or acceptable indexings of the partial recursive functions are sometimes simply called *programming systems.*[2]

Rogers proved two fundamental facts about Gödel numberings. First, he proved that any enumeration of the partial recursive functions which simply has a computable universal function can be effectively translated into any Gödel numbering for the partial recursive functions. (Thus, any two Gödel numberings can each be effectively translated into the other.[3]) Rogers' second theorem proved that from the standpoint of recursion theory all of these standard programming systems are essentially identical. Specifically, building on work by Myhill ([My-55]), he proved that any two Gödel numberings are recursively isomorphic; i.e., that there is a computable one-one, onto map σ such that σ and σ^{-1} effectively calculate the program translations in either direction.

A decade later Blum introduced the notions of abstract complexity measures and size measures for programs ([Bl-67a], [Bl-67b]). Just as Gödel numberings give a useful and fully general abstraction of many salient features of models of computation, Blum measures provide a fully general abstraction of many features of measures of computational complexity such as

---

[2] *Intuitively,* the subscripts, e.g. i and $S_m^n(i, \overline{x})$, represent programs, while x, $\overline{x}$, and $\overline{y}$, are variables representing inputs to the programs. See Section 3.1 of [MY-78] for a simple exposition of acceptable indexings.

[3] This result provides an elegant and very general abstract formulation for results of Church, Kleene, Post, Turing, and others showing that widely different general programming systems, all attempting to calculate broad classes of "effectively computable" functions, all compute exactly the same class of functions. This work of the 1930's and earlier, showing that for many individual examples of programming systems each could compute exactly the same class of functions, is part of the evidence for the *Church-Turing thesis,* which claims that any of a variety of mathematically defined formal systems can exactly capture the *intuitively* understood notion of "effectively computable" function. Thus Rogers' first theorem gives a uniform, abstract, but mathematically precise setting to the basic early work establishing that many individual instance of programming systems can each be translated into the other. If one accepts the Church-Turing thesis, then Rogers work also shows that the acceptable programming systems are all maximal with respect to effective intertranslatability.

time, or space, or tape reversals. In the decade that followed Blum's work, many results and problems in recursion theory were recast in a complexity theoretic light. The isomorphism problems studied by Hartmanis and his students are one of the most important and persistent examples of this. Because Blum's definitions of program size and complexity were strongly tied to the recursion theoretic notion of a Gödel numbering, it was natural to reconsider Rogers' Isomorphism Theorem and other recursion theoretic results in this light. A variety of researchers quickly began work along these lines. Alton, Baker, Berman, Borodin, Constable, Dowd, Hartmanis, Helm, Hopcroft, Kozen, Ladner, Lynch, Machtey, McCreight, Pager, D. Ritchie, Schnorr, Selman, Shay, Winklmann, Young, and others,[4] began research attempting to flush-out the important complexity theoretic issues, including those involved in translation between programming systems. Hartmanis' work was central to this program. While both the syntactic and computational complexity of translations of Gödel numberings was explored by a number of authors following Blum's pioneering work, Hartmanis' work may well have had the broadest impact of all the people working at that time, and much of this work has focused on isomorphism questions.

Isomorphism questions are particularly compelling. If one has a class of functions which one believes is computationally important and fundamental, for example polynomial time computable functions, then programming systems or sets which are isomorphic via functions in this class are computationally *equivalent* with respect to computations in this class. Computational differences among sets or systems which are isomorphic using functions from this class can only be explored by using more restrictive computational bounds. Thus, if one wishes to investigate computational differences among sets or systems which are polynomially isomorphic, as for example one does if they study approximation algorithms for NP-optimization problems, then one must use finer measures of equivalence by restricting the reductions used to some subset of polynomial time reductions, ([CP-90], [Kr-88], [BJY-89]). Thus, polynomial time isomorphisms capture all computational structure in a problem *modulo* fine structure requiring more detailed analysis.

In the remainder of this paper, I have organized Hartmanis' work on isomorphism problems into three sections. First the work on isomorphisms of programming systems. Second the work on isomorphisms of complete sets, including some of the work on one-way functions and collapsing degrees directly motivated by the Berman-Hartmanis conjecture. A final section briefly discusses additional work directly motivated by the Berman-Hartmanis conjecture, including work on sparseness, density, and Kolmogorov complexity.

---

[4]Baker, Berman, Borodin, and Kozen were all students of Hartmanis.

The impact of Hartmanis' work on isomorphism problems, and particularly his ability to articulate important problems and to define general directions of research, has been very broad. No survey such as this can do justice to the full impact of his work, nor do justice to all of the people whose work has been influenced by Hartmanis' work in this area. This is particularly true for someone, such as myself, whose work has been strongly influenced by Hartmanis' papers. Thus, while biases are inevitable in writing a survey such as this, I am very much aware that my own biases enter the survey which follows, and I apologize to the many researchers whose own work has been strongly influenced by Hartmanis' work on isomorphism problems but has been inadequately treated in the sections which follow. And of course, as Juris would surely be the first to acknowledge, his work on isomorphisms has itself been strongly influenced by the work of many colleagues around the world. Thus, to those whose influence has been inadequately traced in the sections which follow, I also apologize.

3.2   The Hartmanis-Baker Conjectures

Hartmanis' first important work on translations of programming systems, ([Ha-74]), was directly motivated by the work of Schnorr. In [Sc-75], Schnorr defined a Gödel numbering to be *optimal* if all Gödel numberings can be translated into the optimal Gödel numbering with at most an additive constant increase in the length of the translated programs. Schnorr showed that all optimal Gödel numberings are isomorphic under linear length bounded mappings, and that, in a certain sense, this isomorphism is just a special case of Rogers' Isomorphism Theorem.[5]

Hartmanis realized that an important question about effective translations concerned not just the "syntactic" complexity of the programs in the source and target languages, but the complexity of the translation process itself. In fact, the stated purpose of [Ha-74] was

> " ... to set up a mathematical model to facilitate the study of quantitative problems about translations between universal languages and to derive ... quantitative results about translations. ... In practical applications one is primarily concerned with two quantitative aspects of translations:
>
> 1. The quality or efficiency of the algorithms translated by σ.
>
> 2. The size and computational complexity of the translator σ itself."

----

[5] The proof that Schnorr gives for this result is actually credited to an anonymous referee.

Building on the recent work of Blum, and in particular Blum's Speed-up Theorem, Hartmanis proved various results showing the impossibility of having translators which always produce absolutely optimal code. He also observed that

" ... under any translation σ ... the running times of the source language algorithms bound recursively the running times of the translated programs and vice versa."

This point was probably well-understood at the time. (Measures with this sort of recursive bound on the running times of translated programs were said to be "recursively related".) Thus in retrospect the most important point in [HA-74] occurs in its conclusion, where there is a careful articulation of some of the general goals of such research:

" ... an important task is to couple, in a very general way, the syntax, semantics, and complexity measures ...
and then (to) restrict somewhat the translations .... How this should be done is not at all clear and seems to be related to the difficult problem of defining 'natural' computational complexity measures."[6]

Hartmanis' realization in [Ha-74] that the critical issue regarding Gödel numberings might not be just the size of the underlying programs, but rather the speed or ease of the translations from one programming system to the next, led to the work that Hartmanis and Baker (then a student

---

[6]This interest in characterizing, perhaps axiomatically, "natural" complexity measures preoccupied a number of people following Blum's pioneering work on "abstract" complexity measures. See e.g., [Al-80], [BCH-69], [C$_i$B-72], [Ha-73], [Pa-69], [Yo-69], [Yo-71a], and [Yo-71b]. These and other authors were convinced that further progress in complexity theory required somehow restricting the "abstract" measures then being considered, but there was then no general agreement about what restrictions might be appropriate. To many, from our current vantage point, this confusion may seem strange. Hartmanis' work on isomorphisms has encouraged us to regard measures which are computationally *similar,* as measured for example by computationally simple isomorphisms or by ease of reductions, as the essentially "natural" measures. Furthermore, much of Hartmanis' work has been tied to specific natural computational models. These two trends in Hartmanis' work have certainly influenced how easily we now view "natural" models of computation and complexity. Focusing on computations in models like polynomial time and log space has been successful because it has tied complexity theory to models which are commonly understood in a wide community. This has allowed the development of the theory to proceed using widely understood concrete natural models, even without having a clear formal, general, notion of what "natural" should really mean. But trying to understand, in a general way, what constitutes a natural complexity measure and programming system still remains a reasonable, if not well-understood and not quite abandoned, problem. (For a variety of recent work in this direction see, e.g., [Ro-87], [Mar-89], [BBJSY-89], and [RC-91].)

of his) did on *simple* Gödel numberings. These are Gödel numberings that any other Gödel numbering can *easily* be translated into. And this, in turn, led to consideration of subrecursive analogues of Rogers' Isomorphism Theorem.

To understand the *ease of translation* problem it is useful to look back at Rogers' 1958 paper. In that paper Rogers gave a standard method for translating from one programming system ϕ' to another system ϕ. Take a fixed universal program $UNIV'$ for ϕ' and an arbitrary S_1^1 function for the system ϕ, and let *univ* be *any* program in the system ϕ which computes the partial recursive function $UNIV'$. Then, since

$$\phi_i'(x) = UNIV'(i,x) = \phi_{S_1^1(univ,i)}(x),$$

any function of the form $\sigma(i) = S_1^1(univ,i)$, computes a translation from the programming system ϕ' to the programming system ϕ. Notice that Rogers' translation shows that *intuitively*, and in reasonable models of computation, the program $\sigma(i)$ simply needs to "preface i onto the input and then turn over control to program *univ*." Seemingly, for a reasonable programming system, the construction of such a program, *univ*, should not be difficult, since it is just an interpreter for the system. Thus, one way to study the complexity of translations from one programming system to another, and hence indirectly to study the complexity of Gödel numberings, is to fix the program *univ* and to study the various ways of computing the S_1^1 function when viewed as a function of it's second argument, i.

In [HB-75], Hartmanis and Baker defined several classes of *simple* Gödel numberings – numberings for which the translations into the simple numbering can always be carried out by simply computable functions. For example, they defined *GNPstfx* to be the class of Gödel numberings such that all translations into these systems can be carried out by a mapping $\sigma(i)$ which simply uses i as a *postfix* to a fixed string of instructions (e.g., "apply the program *univ* to the string obtained by prefixing to the input the following string, i"). Similarly, *GNReg* was defined to be the class of Gödel numberings into which all Gödel numberings can be translated by a mapping $\sigma(i)$ computable by a finite automaton, and *GNLBA* those Gödel numberings such that $\sigma(i)$ is computable by a deterministic linear bounded automaton. With these definitions Hartmanis and Baker proved the proper containments:

$$GNPstfx \subset GNReg \subset GNLBA \subset GNOpt.$$

(Here *GNOpt* are the *optimal* Gödel numberings as defined by Schnorr.)

Hartmanis and Baker next investigated the existence of isomorphisms between programming systems of the same type, showing for example that isomorphisms between Gödel numberings in *GNReg* or in *GNLBA*, respectively, are of bounded computational complexity, but that this is not true for Gödel numberings in *GNOpt*. I.e., there is no recursive bound $t(x)$ such that every two optimal Gödel numberings can each be translated into the other in time t. They also defined three more classes of Gödel numberings, *GNRReg*, the restricted regular Gödel numberings which are given by finite automaton mappings which can only prefix a fixed string or remove a fixed string from the input sequence, *GNPTime* and *GNEXTime*, which are Gödel numberings into which translations can always be obtained by functions computable in time which is, respectively, bounded by a polynomial in the length of the input, and bounded by $2^{c \cdot n}$ for some constant c.[7]

Hartmanis and Baker then set out the first of several complexity bounded isomorphism questions by raising the fundamental question of when isomorphisms within these various complexity-bounded classes of Gödel numberings can be carried out by algorithms within the same complexity class. They conjectured that this could always be done for the three classes *GN-RReg*, *GNPTime* and *GNEXTime*, but they conjectured that for the regular Gödel numberings isomorphisms computable by finite automata would not always exist. For *GNEXTime*, they stated that this closure would follow from a detailed look at the corresponding recursion theoretic proof of Rogers' Isomorphism Theorem.

While a more careful look at Rogers' proof of the recursion theoretic version of the Isomorphism Theorem does *not* in fact yield a proof of the isomorphism result for *GNEXTime*, in [MWY-78] my colleagues Michael Machtey, Karl Winklmann, and I were able not only to provide a proof of this result, but to verify all but perhaps the most important of these isomorphism conjectures (or non-isomorphism in the case of *GNReg*). We could not resolve the conjecture that the class of *GNPTime* Gödel numberings is closed under isomorphisms computable in polynomial time. However in this case we were able to partially confirm the conjecture, showing that between any two polynomial time programming systems there is an isomorphism computable in polynomial time *from a set in* NP.

This remaining conjecture, which should be called the *the Hartmanis-Baker conjecture,* namely, that all polynomial time programming systems are isomorphic under an isomorphism computable in polynomial time remains an open question. (But one should also note the counter-conjecture in [MWY-78] that there are polynomial time programming systems such that any isomorphism between them must be NP-hard.)

In [Ha-82], after working with Leonard Berman in the late 1970's on the question of whether all NP-complete sets are polynomial time isomorphic, Hartmanis returned to the question of whether all polynomial time Gödel numberings are polynomial time isomorphic.

---

[7]WARNING: In [HB-75], Hartmanis and Baker denoted what we have here called *GNEXTime* by *GNPTime*, and they presented no notation at all for what we have here called *GNPTime.*

Most standard Gödel numberings of the partial recursive functions have particularly well-behaved padding functions in the sense that the length of the padded programs can be made to grow in a very regular fashion. Partly because of his work with Berman on the isomorphism question for NP-complete sets, the importance of polynomial time padding functions for proofs of isomorphisms was now clear. For this reason, although he considered various Gödel numberings, in [Ha-82] Hartmanis concentrated, not on *GNPTime* but on a subclass of *GNPTime* which he called *natural* Gödel numberings. These are the programming systems, $\phi \in GNPTime$, into which every other Gödel numbering can be translated via a polynomially computable one-to-one function *whose (left) inverse is also polynomially-time invertible*. Given any of the standard natural Gödel numbering's known to possess a polynomially computable padding function, one can easily induce a polynomially computable padding function for an arbitrary natural Gödel numbering by translating into the standard Gödel numbering, using its padding function, and then translating back. Then, given the fact that the standard Gödel numbering's padding function has a well-behaved rate of growth, Hartmanis carefully performed standard "back-and-forth Schröder-Bernstein" cardinality constructions to induce a polynomial time isomorphism.[8]

Since all natural Gödel numberings are polynomially isomorphic, the question of whether all Gödel numberings in *GNPtime* are *natural* Gödel numberings is equivalent to the Hartmanis-Baker conjecture.

One important feature of [HB-75], and in general all of Hartmanis' work on Gödel numberings, is the recognition that investigating standard *recursion theoretic* properties of programming systems in a complexity theoretic setting is important. This fundamental insight has led in several directions. From the standpoint of this review, the most important direction is the attempt to translate recursion theoretic isomorphism results into complexity-theoretic domains. This has led to fundamental research on when one can expect isomorphisms of the same computational complexity as that of the underlying structures.

Another important direction — one which is beyond the scope of this survey but which nevertheless should at least be mentioned — is toward a consideration of complexity theoretic versions of recursion theoretic programming language constructs, such as S^n_m-theorems, recursion theorems, and composition theorems. In this direction, following the earlier ideas in [Sc-75] and [Ha-75], early work can be found in [MWY-78], but a more careful and more systematic attack has been carried out by other authors, including John Case and his students at Buffalo. Work by Alton, Case,

---

[8] At roughly the same time, Dowd showed that any two r.e. sets which are complete under linear space *m*-reductions are linear space isomorphic, ([Do-82]).

Hamlet, Kozen, Machtey, Riccardi, and Royer following up these initial ideas comes readily to mind. Unfortunately, a discussion of this and related work is beyond the scope of this paper, although some recent work in this area was mentioned earlier, toward the end of footnote 6.

3.3   The Berman-Hartmanis Conjecture

One of the most intensely studied questions in structural complexity theory today is the question of whether all NP-complete sets are polynomially isomorphic. In his paper with Ted Baker on simple Gödel numberings, Hartmanis had asked whether Gödel numberings of a given computational class are always preserved under isomorphisms of the same complexity, and, as we have just seen, the most difficult of these questions proved to be the isomorphism question for polynomial-time Gödel numberings.

In [BH-77] Hartmanis, now working with his student Leonard Berman, turned his attention to similar questions for NP-complete sets and other classes of complete sets. This work was no doubt inspired, not only by Hartmanis' earlier work on Gödel numberings, but also by Myhill's work ([My-55]) on classifying the complete recursively enumerable (r.e.) sets.

Definition 3.3.1 *A set C is* complete *for the class of r.e. sets under many-one reducibilities if for any r.e. set B there is a total recursive function f such that for all x,*

$$x \in B \iff f(x) \in C. \tag{3.1}$$

Myhill showed that all sets which are complete in this fashion (i.e., that are *many-one complete*) are *recursively isomorphic*, and hence from the standpoint of recursion theory are essentially identical. Myhill's proof further characterized the complete r.e. sets by showing that they consist exactly of the *creative* sets.

For Hartmanis and Berman, it was natural to ask at this time whether Myhill's Isomorphism Theorem could be proved in the context of complexity theory. The most important, and perhaps the most obvious place to make the connection was in the class NP of sets recognizable in nondeterministic polynomial time and the class E of sets recognizable in exponential time. In this context, one simply adds to equation (3.1) the requirement that the function f be computable in polynomial time, i.e., that B is reduced to C via a *Karp*-reduction.

At the time it was widely believed that it was fruitful to consider the relation between P and NP to be roughly analogous to the relation between the recursive sets and the r.e. sets. If this analogy were to hold, then from Myhill's work it would be natural to believe that all NP-complete sets should be polynomially isomorphic. Furthermore, in his 1977 doctoral dissertation, Berman, ([Be-77]), proved that all sets which are complete for E are interreducible via one-one *length increasing* polynomially computable reductions.[9]

Given this history, it was natural for Berman and Hartmanis to conjecture that all sets which are complete for NP under Karp reducibilities should be isomorphic under polynomial time reductions.

In recursion theory, Myhill's proof that all many-one complete sets are recursively isomorphic splits into two parts. As described in more detail in the following pages, one first proves that all many-one complete sets are *creative* and hence that they all have *padding functions*. Since one does not have to worry about the number of steps in the computation, the proof that any two sets which are mutually many-one reducible and which both have padding functions are isomorphic is then a simple adaptation of the standard Schröder-Bernstein proof that any two sets which are mutually one-one interreducible have the same cardinality.

In [BH-77], Berman and Hartmanis showed that all *known* examples of NP-complete sets possess padding functions, and furthermore that the standard examples of NP-complete sets possess padding functions which satisfy a nice growth requirement. Several definitions of polynomial time padding functions are possible. I give here a definition which follows the standard definitions from recursion theory. This definition can be shown ([MY-85]) to be equivalent to those employed by Hartmanis.

Definition 3.3.2 *A polynomial time padding function for a set A is a one-one polynomial time computable function p of two arguments such that for all x and y,*

$$p(x, y) \in A \iff x \in A,$$

and p is invertible in polynomial time.

Because a set, A, which has a polynomial time padding function is polynomial time isomorphic to the cylinder set $A \times N$, such sets are called (polynomial) *cylinders*. If A is polynomial time paddable with padding function p and if f many-one reduces B to A in polynomial time, then

$$x \in B \iff p(f(x), x) \in A.$$

Thus *any* such set B is *one-one* reducible to A in polynomial time *via a function which is invertible* in polynomial time. So if we have two sets A

---

[9]I.e., that they are *pseudo p-isomorphic* in the sense later used by Watanabe ([Wa-85]) in another proof of this and similar results. A somewhat simpler proof of this result has recently been obtained by Ganesan and Homer, ([GH-88]), who also prove the result for one-one functions (but not for *length increasing* one-one functions) for *nondeterministic* exponential time.

and B which are many-one polynomial time interreducible, and if each has a padding function, then A and B are one-one interreducible in polynomial time. By using standard, Schröder-Bernstein type constructions, one can use the map from A to B and the map from B to A to produce a one-one, onto map from B to A, i.e., a polynomial time *isomorphism* whose inverse obviously also reduces A to B in polynomial time.

From a complexity theoretic point of view, the problem with the most obvious form of this construction is that the process which constructs the isomorphism goes back and forth between the reduction from A to B and the reduction from B to A to build the isomorphism, and in doing so keeps track of the values computed on all earlier elements. Consequently, this construction may have an exponential number of elements to examine. Hence, even when the original reductions and the padding functions are polynomial time computable, the most obvious isomorphism may only be computable in exponential time.

However, as we mentioned above, Hartmanis and Berman observed two things: first, that all of the standard NP-complete sets known at that time had polynomial time padding functions, and second, that for many of these it is also easy to obtain a padding function which has a very nice rate of growth. Typically

$$|p(x,y)| \quad > \quad |x|^2 + 1,$$

or sometimes simply strictly size increasing.

Given two polynomially interreducible cylinders, one of which has a polynomial time padding function with a suitable rate of growth, Hartmanis and Berman were able to show by careful counting in the Schröder-Bernstein construction that the isomorphism between the two sets can be computed in polynomial time.[10] As remarked in the preceding section, this same technique was later used by Hartmanis in [Ha-82] to prove his polynomial isomorphism result for *natural* Gödel numberings.

Thus, perhaps the most important result in [BH-77] is the realization that all sets then known to be NP-complete are polynomially isomorphic, i.e., are identical up to a polynomially computable permutation of their elements. However, in spite of the fact that all sets then known to be NP-complete could be shown to be polynomially isomorphic, the result did not follow from the definition of NP-complete sets, but instead relied on the fact that all known examples of NP-complete sets could easily be seen to have polynomial time padding functions, and so the general question remained unanswered.

----

[10]It has been shown in [MY-85] that *no* requirement at all on the rate of growth of the padding functions is necessary. This is an improvement in the proof only and not the results, since in practice one always easily sees that one of the padding functions *does* have a suitable growth rate.

Perhaps encouraged by their success with all known NP-complete sets and relying also on the obvious analogy with Myhill's recursion theoretic results, Hartmanis and Berman went on to state, ([BH-77]), *the Berman-Hartmanis conjecture: all NP-complete sets are isomorphic under polynomial time computable permutations*. This conjecture remains open, as does the corresponding question for E and the question of whether all NP-complete sets are pseudo *p*-isomorphic.

As we shall see, the Berman-Hartmanis conjecture has motivated much of the past decade's important research in structural complexity theory, and it is probably fair to say that it was widely accepted at least until it was challenged in the mid-1980's.

It was also observed in [BH-77] that the technique of using padding functions applies also to sets in polynomial space, thus showing the existence of polynomial *time* computable isomorphisms between all *known* PSPACE-complete problems. Similarly, this same technique was used again in [HB-78] to establish polynomial time isomorphisms for additional problems complete for NP, for PSPACE, and for ESPACE.

Next Hartmanis considered space bounded reducibilities, and he observed that the padding technique also works for log-space computable reductions. In [Ha-78b], he applied this technique to prove the isomorphism of various complete problems for NP, P, PSPACE, nondeterministic log-space (NL), and context sensitive languages (CSL), under log-space computable reductions. In [HIM-78], *1-L* reductions (logspace with a one-way read head) were considered and various completeness notions were proven, including constructions of sets which are complete for NL under *logspace* reductions but *not* complete (and hence not isomorphic to standard complete sets) *under 1-L reductions*. Hartmanis discussed many of these results for sets complete under log-space, as well as polynomial-space reductions in Sections 2 and 3 of [Ha-78a].[11]

More recently, Hartmanis and Mahaney, ([HM-81]), have shown that no sparse set can be hard for NL under *1-L* reductions, and Allender, ([Al-88]), has proved that all sets complete for PSPACE and for exponential time under *1-L* reductions are polynomial time isomorphic, and Ganesan, ([Ga-90]), has a similar result for nondeterministic exponential time.

In his paper on natural Gödel numberings ([Ha-82]), Hartmanis introduced the notion of a *natural creative set*. We have already mentioned that *creative sets* play a key role in Myhill's proof that all sets complete for the r.e. sets are recursively isomorphic. Basically, Myhill proved that a set is

---

[11]Dowd, in interesting unpublished but widely circulated manuscripts, ([Do-78&82]), defined and attacked isomorphism problems for a number of interesting complexity classes and proved isomorphism and reduction theorems paralleling many of these results.

complete *if and only if* it is creative and then proved that all creative sets are cylinders.

Definition 3.3.3 *A r.e. set C is* r.e.-creative *if there is a total recursive function f such that for every r.e. set W_i,*

$$W_i \subseteq \overline{C} \implies f(i) \in \overline{C} - W_i.$$

The idea here is that $\overline{C}$ cannot be r.e., because given any candidate W_i for being $\overline{C}$, f *produces* an element $f(i)$ which witnesses that $\overline{C}$ is not a subset of W_i. Hence f is called a *productive* function for $\overline{C}$.

Perhaps anticipating the importance of creative sets for research in complexity theory, Hartmanis defined a creative set to be a *natural* creative set if the productive function f can be taken one-one and polynomially computable with a polynomial time computable inverse. He defined a r.e. set to be a *natural* complete set if every r.e. set can be reduced to it via such a polynomially computable and invertible function, and he proved that all natural complete sets are polynomially isomorphic.[12] He also showed that the natural complete sets are exactly the natural creative sets, and thus showed that all natural creative sets are polynomially isomorphic.[13]

With this background, it is natural to ask whether the same ideas can be used to investigate NP-complete sets. In view of Hartmanis' work on r.e. sets, the natural thing to do for NP-complete sets is to require, as was done in [Ha-82], that the productive functions range over all polynomial-time computable one-one functions with polynomially computable inverses, to require that the set C be in NP, and to let the sets W_i range over all sets in NP. Using essentially this definition, in 1981 Ko and Moore ([KM-81]) proved that no NP-creative sets exist. In fact they proved that no such productive functions could exist for any sets in exponential time.

Returning to the original definition of *r.e.-creative* sets given above, in [JY-85] Deborah Joseph and I defined *k-creative* sets for NP by requiring only that the productive functions be one-one and computable in polynomial time (not necessarily invertible) and that the sets W_i range, not over *all* NP-complete sets, but only over sets accepted in nondeterministic time bounded by a polynomial of degree k.

With this definition we were able to show that every k-creative set is NP-complete and furthermore that *every* one-one polynomially computable and

---

[12]The corresponding isomorphism question for *polynomially*-complete sets, i.e., those r.e. sets such that all r.e. sets can be reduced to them in polynomial time, is an important open question which sits between the Hartmanis-Baker conjecture and the Berman-Hartmanis conjecture.

[13]This characterization of the natural creative sets is actually credited to an anonymous referee.

honest function is a productive function for some k-creative set. This defined a whole new class of NP-complete sets, and the only obvious way to prove these sets polynomially isomorphic to more standard NP-complete sets is to have the productive functions have inverses which are computable in polynomial time.[14] [JY-85] *thus specifically conjectured that if one-way functions exist, then not all k-creative sets are polynomial isomorphic to SAT, and thus that the Berman-Hartmanis conjecture fails.* Others, including Osamu Watanabe ([Wa-87a]), gave results pointing out that if f is a one-way function and if C is any NP-complete set, then $f(C)$ is an NP-complete set for which there is no apparent reason to believe that it is polynomial time isomorphic to C. Later [KMR-86] conjectured the converse of the Joseph-Young conjecture: essentially Kurtz, Mahaney and Royer conjectured that if one-way functions fail to exist, then the Berman-Hartmanis conjecture is true.

Perhaps partly as a result of these counter-conjectures, there has been intense interest in the Berman-Hartmanis conjecture in the past few years. One such line of work has focused on the question of how the existence of one-way functions either explicitly or implicitly influences collapsing or noncollapsing of various polynomial many-one degrees. Since the Berman-Hartmanis conjecture implies that the collection of sets polynomially many-one equivalent to SAT, (i.e., the polynomial many-one *degree* for NP), *collapses* to a single isomorphism type, and since no nontrivial examples of such collapsing degrees were known at the time of the conjectures, another line of work has centered on the search for reasonably computationally easy (say at or below exponential time) places where polynomial many-one degrees collapse to a single isomorphism type.

It is beyond the scope of this paper to give a careful survey of this latter work. Furthermore, given the very well-written, fair, and extensive survey of work on collapsing degrees given in the introduction to [KMR-86], given at the 1987 Structures in Complexity Theory Conference ([KMR-87b]), and given in the tutorial in this book, ([KMR-90]), it would be presumptuous to give such a discussion here. I strongly encourage the reader to read one of the surveys or the tutorial by Kurtz, Mahaney and Royer.

---

[14]In [Ho-86], Steve Homer, observing that general computational models like Turing machines can be embedded in tiling problems, embedded the construction of k-creative sets into tiling problems, thus providing "combinatorial" examples of k-creative sets. More recently, Wang ([Wa-89]) has revisited definitions of polynomial creativity. He has shown that the Ko-Moore negative results are sensitive to the indexings used for the polynomial bounds, and with reasonable indexings has obtained p-creative sets in exponential time. In general, Wang gives a detailed analysis of various forms of polynomial creativity which sit between Ko and Moore's notions of polynomial creativity and the k-creative sets of Joseph and Young. Ganesan, ([Ga-90]), shows that it is possible for k-creative sets to have one-way productive functions while remaining isomorphic to SAT, but this does *not* refute the Joseph-Young conjectures of [JY-85]).

Here, I will mention only a few of the results discussed there. First, the very nice results of Ko, Long, and Du, ([KLD-86]), showing that if one-way functions exist, then sets that are interreducible under one-one, polynomial time computable and size increasing functions need not collapse to a single-isomorphism type. Indeed, for these degrees, noncollapsing to a single isomorphism type is equivalent to the existence of one-way functions. Second the very beautiful construction by Kurtz, Mahaney and Royer, ([KMR-86]) showing (without assumptions) that there is a many-one degree in exponential time which *does* collapse to a single polynomial isomorphism type.

Returning to the conjecture of Kurtz, Mahaney, and Royer that nonexistence of one-way functions implies the collapse of the NP-complete sets to a single polynomial isomorphism type, Hartmanis and his student Lane Hemachandra have recently constructed oracles relative to which one-way functions do *not* exist, P $\neq$ NP, and yet the NP-complete sets do not collapse to a single isomorphism type (nor even a single primitive recursive isomorphism type!), ([HH-87]).

It would be interesting to know whether there is an oracle relative to which the Joseph-Young conjecture that the existence of one-way functions implies that the NP-complete sets do *not* collapse to a single polynomial isomorphism type *fails*. In 1987, Hartmanis and Hemachandra pointed out ([HH-87]), "the nearest result to this is a recent theorem of Goldsmith and Joseph ([GJ-86]) that constructs an oracle A for which all $\leq_m^p$-complete sets for NP^A are p^A-isomorphic."[15]

Recently, Homer and Selman, ([HS-89]), have constructed an oracle relative to which all Σ_2^P-complete sets collapse to a single polynomial isomorphism type while P $\neq$ NP. Relative to this oracle, one-way functions do not exist (UP = P), and EXP collapses to Σ_2^P. Furthermore, relative to this oracle, NP does not have any polynomially inseparable sets, and this together with UP = P shows that relative to this oracle, although NP $\neq$ P, public key cryptography is not possible. Since it is also possible, using techniques due to Kurtz, to construct oracles relative to which *not* all Σ_2^P-complete sets collapse to a single polynomial isomorphism type, the complete sets for Σ_2^P are thus the first known example of a "natural" complexity class for which the polynomially many-one complete sets can be made either to collapse to a single isomorphism type or not to collapse to a single isomorphism type by choice of a suitable oracle.

Long, ([Lo-88]), building on the unrelativized results in [KLD-86], and responding partly to questions in [KMR-87b] has proven that there is a sparse oracle, A, relative to which E $\neq$ NP, a one-way function f^A exists,

---

[15]For details beyond those of [GJ-86], including those which guarantee $NP^A \neq P^A$, the reader should consult ([Go-88]).

and there is a *2-tt*-complete set, C^A, for NP^A for which C^A and $f^A(C^A)$ are equivalent under length-increasing one-one polynomial time reductions but C^A and $f^A(C^A)$ are not polynomially isomorphic.[16]

In even more recent work, ([KMR-89]), Kurtz, Mahaney, and Royer have proven that, with respect to random oracles, the Berman-Hartmanis conjecture fails with probability one. Like the work of Long just discussed, this work also intimately ties the failure of the isomorphism question to the existence of one-way functions, and in doing so supercedes Long's work announced in [Lo-88]. In [FKR-89], Fenner, Kurtz, and Royer have also proved the very nice result that P = PSPACE if and only if *every* two sets which are one-one interreducible in polynomial time are polynomial time isomorphic.

Before closing this section, I cannot resist mentioning work of Steve Mahaney and I, ([MY-85]), which shows that, if the Berman-Hartmanis conjecture fails, then it fails very badly indeed. In fact, building on earlier work in [Ma-81] and [Yo-66], Mahaney and I showed that *any* polynomial many-one degree which fails to collapse to a single polynomial isomorphism type splits into an infinity of isomorphism types which are so fractured under one-one length increasing polynomial reductions that one can embed *any* countable partial order into the ordering by one-one length increasing polynomial reductions.

In summary, the realization in 1976 that all *known*, and presumably all important and naturally occurring, NP-complete sets are polynomially isomorphic, and the resulting Berman-Hartmanis conjecture that *all* NP-complete sets are polynomial time isomorphic has opened one of the most fruitful areas of research in structural complexity theory of the past dozen years. Juris Hartmanis and his students not only initiated research on this important topic, but Hartmanis himself remains one of the most important contributors to the area.

3.4   Sparseness, Density, Kolmogorov Complexity

One test of the impact of a scientific hypothesis is the extent to which it leads to important related work. Hartmanis' work on isomorphisms, and particularly his work on the Berman-Hartmanis conjecture meets this criterion.

---

[16]Kurtz, Mahaney, and Royer had already shown that it is possible to construct sparse oracles A relative to which $E^A = NP^A$ and their various collapsing and noncollapsing degree constructions, ([KMR-86]), work to get collapsing and noncollapsing degrees which are *2-tt*-complete for NP^A. By keeping $E^A \neq NP^A$, Long's construction seems to require the existence of one-way functions, and so has more relevance to the conjecture discussed here.

In the preceding section, we discussed two related sets of research ideas spawned by this conjecture. We saw there that one way to refute the Berman-Hartmanis conjecture might be to tie it closely to the existence of one-way functions, and we saw that this line of attack has led to important research making exactly such connections, as well as to important new research on collapsing degrees.

Another line of attack on the Berman-Hartmanis conjecture leads to questions about the density and sparseness of sets complete for NP and classes. The importance of this approach was realized by Berman and Hartmanis already in 1976, and results about the nonexistence of sparse complete sets for various complexity classes was an important part of [HB-76]:

> " ... if all NP-complete sets are p-isomorphic, then they all must have similar densities and, for example, no language over a single letter alphabet can be NP-complete. We show that complete sets in E and ESPACE cannot be sparse and therefore they cannot be over a single alphabet. Similarly, we show that the hardest context-sensitive languages cannot be sparse. We also relate the existence of sparse complete sets to the existence of simple combinatorial circuits ... "

From the above, and from the fact that P = NP *would* imply the existence of sparse NP-complete sets, it was widely believed that a proof that no NP-complete set could be sparse would constitute confirming evidence for the Berman-Hartmanis conjecture. In addition to its relevance to the isomorphism question, Hartmanis motivated research on sparse sets by pointing out that it addresses the question of whether complete sets are hard because they have a lot of hard instances or whether they are hard simply because they have infinitely many, but widely scattered, instances. It was also known at this time that any set reducible to a sparse set could be solved (perhaps nonuniformly) by polynomial size circuits. Thus the conjecture in [HB-76] and [BH-77a] that no sparse set could be NP-complete was the source of exciting research in the late 1970's. This conjecture on sparse sets was confirmed by Steve Mahaney, once more a student of Hartmanis. A very readable account of the genesis of Mahaney's proof was provided in [HM-80],[17] and we have used this source in preparing our own brief summary.

Perhaps the first published attack on the Berman-Hartmanis conjecture was by Book, Wrathall, Selman, and Dobkin who related the no-sparse-complete-set conjecture to the still open question of whether any tally set could be *inclusion*-complete for NP, ([BWSD-77]).

---

[17]But ignore their Corollary 3!

The first significant progress on this no-sparse-complete-set conjecture was by Piotr Berman, ([Be-78]), who proved that if there is a polynomial time function, (i.e., a *Karp reduction*), with a sparse range which reduces an NP-complete set to any set A, then P = NP. Of course it follows that no tally set can be complete for NP unless P = NP.

Steve Fortune, quickly modified Piotr Berman's proof to obtain a proof that complete sets for coNP cannot be sparse and cannot be reduced to sparse sets, ([Fo-79]). And then Steve Mahaney, in his doctoral dissertation, provided a very pretty proof, nonconstructively utilizing census functions which give counts of the exact number of elements in any sparse set, that no NP-complete set can be sparse or can be reduced to a sparse set, ([Ma-82]).

We might remark in passing that, while Mahaney proved that no set which is NP-complete under Karp reductions can be sparse or can be reduced to a sparse set under Karp reductions, and thus answered the conjecture on sparse complete sets set by Hartmanis and Berman, it still remains an open question whether polynomially sparse sets can be complete for NP under more general *Cook reductions*, i.e., under polynomial time *Turing* reductions.[18]

Various people have worked on this problem. Karp and Lipton and Sipser, ([KL-80]), Long, ([Lo-82]), and Mahaney, ([Ma-82]), all show that the existence of various forms of sparse *hard* or *complete* sets for NP imply the collapse of the polynomial time hierarchy. For example, perhaps the strongest of these results is Long's, which shows that if SAT is Cook reducible to a sparse set, then every set in the polynomial time hierarchy is reducible to the *same* sparse set. If follows, for example, that if there are *either* sparse or cosparse sets which are Cook-complete for NP, then the polynomial time hierarchy collapses to Δ_2^P.

While the original no-sparse-complete-set conjecture of Hartmanis and Berman was answered by Mahaney, the line of research spawned by this conjecture on sparseness is still important.[19] As one example we once again quote Hartmanis, ([Ha-87b]):

> "......... the proof that E $\neq$ NE iff there exist sparse sets in NP $-$ P, ([HIS-85]), uses the census function and so did Jim

---

[18]The problem of whether Cook reductions and Karp reductions differ on NP remains an open question. A recent discussion of this problem, as well as some partial solutions, have been given by Longpré and Young, ([LY-88]). Longpré is again a student of Hartmanis.

[19]An interesting survey of selected work on sparse sets, tally sets, and Kolmogorov complexity, can be found in [Bo-88]. Book and his colleagues have long studied sparse sets and tally sets. Book's survey cites [BH-78], [Be-78], [Fo-79], Ma-82], as well as [KL-80] for providing inspiration for much of the work surveyed there. Mahaney's [Ma-89] is a useful survey of many of the same topics from a somewhat different perspective.

Kadin's optimal collapse of PH to $P^{SAT[O(log(n))]}$ under the assumption that there exist sparse S in NP such that NP $\subseteq P^S$, ([Ka-88], [Ha-87a]). It was the proof of this last result which inspired Schöning and Wagner ([SW-88]) to derive their simple proof of ... various collapses of hierarchies, some of which were superseded by Immerman's closure result ([Im-88])."

Hartmanis is referring here to the very nice new result of Szelepcsényi and Immerman, ([Sz-88], [Im-88], [HR-88]), which shows that various nondeterministic space classes are closed under complementation. (And we note that Immerman is a former student of Hartmanis!)

The preceding results on density and sparseness trace one research thread in a program attempting to *prove* the Berman-Hartmanis conjecture. But Hartmanis has also worked on *disproofs* of the conjecture. Following an oracle construction by Kurtz providing an oracle relative to which the isomorphism conjecture fails and following the work by Joseph and Young constructing k-creative sets which seem not to be polynomially isomorphic to standard NP-complete sets, Hartmanis started a new line of attack, using *Kolmogorov complexity* to try to deduce evidence that the Berman-Hartmanis conjecture *fails*. Informally, he proved, ([Ha-83]), that "if the satisfiability of Boolean formulas of low Kolmogorov complexity can be determined in polynomial time, then there exist NP-complete sets which are not p-isomorphic to SAT."

Here Hartmanis had in mind a "generalized Kolmogorov complexity of finite strings which measures *how far* and *how fast* a string can be compressed." To this end, following similar usage and definitions by Levin ([Le-73] and [Le-84]), Ko ([Ko-83], and Sipser ([Si-83]), Hartmanis took a universal Turing machine M_u and functions g and G mapping natural numbers into natural numbers and defined a generalized Kolmogorov complexity of strings x by

Definition 3.4.1 $K_u[g(n), G(n)] = \{x \mid (\exists y)[|y| \leq g(|x|)$ *and* $M_u(y) = x$ *in fewer than* $G(|x|)$ *steps*$\}$.

"Thus the function g measures by how much the string x must be compressed and G measures how much time is needed to compute x from y."

Using these notions, in [Ha-83] Hartmanis considered sets of small generalized Kolmogorov complexity and proved

Theorem 3.4.2 *If there exists a set S_0 in P such that*

$$K[log(n), n^{log(n)}] \cap SAT \subseteq S_0 \subseteq SAT,$$

then $SAT - S_0$ is an NP-complete set which is not p-isomorphic to SAT.

However even Hartmanis has pointed out that this approach is unlikely to disprove the Berman-Hartmanis conjecture since, as he shows later in the paper, the necessary hypothesis implies that E = NE. and probably few people believe that E = NE. Nevertheless, the approach will surely lead to interesting work. For example, Deborah Joseph has conjectured that $K[log(n), n^{log(n)}] \cap SAT$ and $\overline{SAT}$ are actually polynomially *inseparable*.[20] This conjecture suggests an interesting interplay between generalized Kolmogorov complexity and polynomial separability and inseparability.[21]

With this latest work on Kolmogorov complexity, Hartmanis has once again helped launch a new and promising line of research in structural complexity theory. My hope is that this overview of research on isomorphism problems has helped elucidate how effective Juris Hartmanis has been in articulating new directions for research. Thus this recent example of Hartmanis' use of Kolmogorov complexity to attack the isomorphism problem would seem seem an appropriate place to close my survey.

However, before closing I'd like to observe that Hartmanis' recent work on Kolmogorov complexity circles back to a research topic that predates his work on the isomorphism question. Kolmogorov complexity is concerned with succinct representation of data, and Hartmanis has had a long term interest in this topic. An important theme in his earlier work, and one which continues today, is his interest in succinctness of programming representations, ([Ha-63a], [Ha-63b], [BH-79], [Ha-80], [Ha-81], and [Ha-83b]). While it is beyond the scope of this paper to explore the work on succinctness, it should be noted that Hartmanis' recent work on Kolmogorov complexity reflects a long-term intellectual interest in bringing a wide variety of ideas to bear on problems in structural complexity theory.

Acknowledgements: The author would like to thank Eric Allender, Deborah Joseph, Tim Long, Steve Mahaney, and Jim Royer for reading the initial version of this paper and suggesting helpful changes and additions.

---

[20] Juris Hartmanis, private communication.

[21] [LV-90] is a very nice survey of work on Kolmogorov complexity by Li and Vitányi, and includes a good discussion of Hartmanis' contributions. (Ming Li was again a student of Hartmanis.)

3.5 Bibliography

The papers listed below have been grouped according to topic. The three major topics listed correspond to the major sections of this paper: the early work on the isomorphism problem for Gödel numberings (Section 2 of this paper), the work directly related to the Berman-Hartmanis conjecture (Section 3), and more recent work on sparse sets and Kolmogorov complexity directed toward proving or disproving the Berman-Hartmanis conjecture (the work surveyed in Section 4). Within each of these categories I have begun by listing Hartmanis' papers chronologically, with an obviously incomplete attempt made to reference conference papers together with the corresponding archival journal paper. This is followed by a list of related papers by other authors, organized alphabetically. I hope that from this list of references one gains some additional understanding of how Hartmanis' work has interacted with that of other researchers and of the impact his work has had on this branch of structural complexity theory.

3.5.1 INTRODUCTION AND GENERAL BACKGROUND

Primary Papers - J. Hartmanis

[HH-71] Hartmanis, J. and J. Hopcroft, "An overview of the theory of computational complexity," *J Assoc Comp Mach* **18** (1971), 444-475.

[Ha-78a] Hartmanis, J., *Feasible Computations and Provable Complexity Properties*, SIAM CBMS-NSF Regional Conference Series in Applied Math **30** (1978).

[Ha-81] Hartmanis, J., "Observations About the Development of Theoretical Computer Science," *Annals History Comput* **3** (1981), 42-51.

Related Papers - Other Authors

[BBJSY-89] Bertoni, A., D. Bruschi, D. Joseph, M. Sitharam, and P. Young, "Generalized Boolean hierarchies and Boolean hierarchies over RP," final version available as Univ Wisc CS Tech Report; short abstract in *Proc 7th Symp Foundations Computing Theory*, (FCT-1989), *Springer-Verlag, LNCS* **380**, 35-46.

[Bl-67a] Blum, M., "A machine-independent theory of complexity of recursive functions," *J Assoc Comput Mach* **14** (1967), 322-336.

[Bl-67b] Blum, M., "On the size of machines," *Inform and Control* **11** (1967), 257-265.

[Bo-69] Borodin, A., "Complexity classes of recursive functions and the existence of complexity gaps," *Proc 1st ACM Symp Theory Comput* (1969), 67-78.

[BCH-72] Borodin A., R. Constable and J. Hopcroft, "Dense and nondense families of complexity classes," *Proc 10th IEEE Switching and Automata Theory Conf,* (1969), 7-19.

[BJY-89] Bruschi, D., D. Joseph, and P. Young, "A structural overview of NP optimization problems," *Proc 2nd International Conference on Optimal Algorithms; Springer-Verlag, Lecture Notes Comp Sc,* (1989), 27 pages, to appear.

[Co-69] Constable, R., "The operator gap," *Proc 10th IEEE Symp Switching Automata Theory,* (1969), 20-26.

[Co-70] Constable, R., "On the size of programs in subrecursive formalisms," *Proc 2nd ACM Symp Theory Comput* (1970), 1-9.

[CB-72] Constable, R. and A. Borodin, "Subrecursive programming languages, Part I: Efficiency and program structure," *J Assoc Comput Mach* **19** (1972), 526-568. Some of the results in this paper first appeared in, "On the efficiency of programs in subrecursive formalisms," presented at the 11th IEEE Symp Switching and Automata Theory, (1970), 60-67.

[CP-89] Crescenzi, P. and Panconesi, A., "Completeness in approximation classes," *Proc 7th Symp Foundations Computing Theory,* (FCT-1989), *Springer-Verlag, LNCS* **380**; final version to appear in *Inform and Comput.*

[Kr-88] Krentel, M. "The complexity of optimization problems," *J Comput System Sci* **36** (1988), 490–509; (preliminary version in *Proc 18th ACM Symp Theory Comput,* (1986), 69–76.)

[La-75] Ladner, R., "On the structure of polynomial time reducibility," *J Assoc Comput Mach* **22** (1975), 155-171.

[LLS-75] Ladner, R., A. Selman, and N. Lynch, "A comparison of polynomial time reducibilities," *Theor Comput Sci* (1975) 103-123.

[MY-78] Machtey, M., and P. Young, "An Introduction to the General Theory of Algorithms," *Elsevier North Holland,* New York (1978), 1-264.

[Mar-89] Marcoux, Y., "Composition is almost as good as *S-1-1*," *Proc 4th IEEE Symp Structure Complexity Theory,* (1989), 77-86.

[MM-69] McCreight, E. and A. Meyer, "Classes of computable functions defined by bounds on computation," *Proc 1st ACM Symp Theory Computing,* (1969), 79-81.

[Me-72] Meyer, A., "Program size in restricted programming languages," *Inform and Control* **21** (1972), 382-394.

[Ri-68] Ritchie, D., *Program Structure and Computational Complexity,* Ph.D. Thesis Harvard University (1968).

[Ro-87] Royer, J., "A Connotational Theory of Program Structure," *Springer Verlag LNCS* **273,** (1987).

[RC-91] Royer, J. and A. Case, "Intensional Subrecursion and Complexity Theory," *Research Notes in Theoretical Computer Science,* to appear.

[St-87] Stockmeyer, L., "Classifying the computational complexity of problems," *J Symb Logic* **52** (1987), 1-43.

[Yo-69] Young, P., "Toward a theory of enumerations," *J Assoc Comput Mach* **16** (1969), 328-348.

3.5.2   ISOMORPHISMS OF GÖDEL NUMBERINGS

Primary Papers - J. Hartmanis

[Ha-74] Hartmanis, J., "Computational complexity of formal translations," *Math Systems Theory* **8** (1974), 156-166. Work presented in this paper concerning formal models for translation extends some of the work in [CH-71], "Complexity of formal translations and speed-up results," (with R. Constable) presented at *STOC* 1971.

[HB-75] Hartmanis, J. and T. Baker, "On simple Gödel numberings and translations," *SIAM J Comput* **4** (1975), 1-11. This paper extends work in "On simple Gödel numberings and translations," presented at *ICALP* 1974.

[Ha-82] Hartmanis, J., "A note on natural complete sets and Gödel numberings," *Theor Comput Sci* **17** (1982), 75-89.

Related Papers - J. Hartmanis

[Ha-73] Hartmanis, J., "On the problem of finding natural computational complexity measures," *Proc Symp Math Found Comput Sci,* (1973), 95-103.

[Ha-81] Hartmanis, J., "Observations about the development of theoretical computer science," *Annals History of Comput* **3** (1981), 42-51. The work in this paper was originally presented at *FOCS* (1979).

Related Papers - Other Authors

[Al-76] Alton, D., "Nonexistence of program optimizers in several abstract settings," *J Comput System Sci* **12** (1976), 368-393.

[Al-77a] Alton, D., "Program structure, 'natural' complexity measures, and subrecursive programming languages," *Proc* **2**nd *Hungarian Comput Sci Conf*, Akad Kiado, Budapest, (1977).

[Al-77b] Alton, D., " 'Natural' complexity measures and time versus memory: some definitional proposals," *Proc* **4**th *ICALP, Springer Verlag LNCS* **52,** (1977).

[Al-80] Alton, D., " 'Natural' complexity measures and subrecursive complexity," in *Recursion Theory: Its Generalizations and Applications,* editors F.R. Drake and S.S. Wainer, Cambridge Univ Press, (1980), 248-285.

[HMY-73] Helm, J., A. Meyer, and P. Young, "On orders of translations and enumerations," *Pacific J Math* **46** (1973), 185-195.

[Ko-80] Kozen, D., "Indexings of subrecursive classes," *Theor Comput Sci* **11** (1980), 277-301.

[MWY-78] Machtey, M., K. Winklmann, and P. Young, "Simple Gödel numberings, isomorphisms, and programming properties," *SIAM J Comput* **7** (1978), 39-60. This paper extends work in, "Simple Gödel numberings, translations and the P-hierarchy," presented at *STOC 1976*.

[Pa-69] Pager, D., "On finding programs of minimal length," *Infor and Control* **15** (1969), 550-554.

[Ri-81] Riccardi, G., "The independence of control structures in abstract programming systems," *J Comput System Sci* **22** (1981), 107-143.

[Ro-58] Rogers, H., "Gödel numberings of partial recursive functions," *J Symbolic Logic* **23** (1958), 331-341.

[Sc-75] Schnorr, C. P., "Optimal enumerations and optimal Gödel numberings," *Math Systems Theory,* **8** (1975), 182-191.

[SY-78] Shay, M. and P. Young, "Characterizing the orders changed by program translators," *Pacific J Math* **76** (1978), 485-490.

[Yo-69] Young, P., "Toward a theory of enumerations," *J Assoc Comput Mach* **16** (1969), 328-348.

[Yo-71a] Young, P., "A note on 'axioms' for computational complexity and computation of finite functions," *Inform and Control* **19** (1971), 377-386.

[Yo-71b] Young, P., "A note on dense and nondense families of complexity classes," *Math Systems Theory* **5** (1971), 66-70.

3.5.3   THE BERMAN-HARTMANIS CONJECTURE

Primary Papers - J. Hartmanis

[BH-77a] Berman, L. and J. Hartmanis, "On isomorphism and density of NP and other complete sets," *SIAM J Comput* **6** (1977), 305-322. The results in this paper were first presented in [HB-76], "On isomorphism and density of NP and other complete sets," at *STOC* (1976).

[BH-77b] Berman, L. and J. Hartmanis, "On polynomial time isomorphisms of complete sets," *Proc 3rd Theor Comput Sci GI Conf, Springer Verlag LNCS* **48** (1977), 1-15.

[HB-78] Hartmanis, J. and L. Berman, "On polynomial time isomorphisms of some new complete sets," *J Comput System Sci* **16** (1978), 418-422.

Related Papers - J. Hartmanis

[Ha-78a] Hartmanis, J., *Feasible computations and provable complexity properties, SIAM CBMS-NSF Regional Conf Series in Applied Math* **30**, (1978).

[HIM-78] Hartmanis, J., N. Immerman and S. Mahaney, "One-way, log-tape reductions," *Proc 19th IEEE FOCS*, (1978), 65-72.

[Ha-78b] Hartmanis, J., "On log-tape isomorphisms of complete sets," *Theor Comput Sci* **7** (1978), 273-286.

[HM-81] Hartmanis, J. and S. Mahaney, "Languages simultaneously complete for one-way and two-way log-tape automata," *SIAM J Comput* **10** (1981), 383-390.

[Ha-82] Hartmanis, J., "A note on natural complete sets and Gödel numberings," *Theor Comput Sci* **17** (1982), 75-89.

[Ha-83a] Hartmanis, J., "Sparse sets in $NP - P$," *Infor Processing Letters,* **16** (1983), 55-60.

[HH-87] Hartmanis, J. and L. Hemachandra, "One-way functions, robustness, and the non-isomorphism of NP-complete sets," *Proc 2nd IEEE Symp Structure Complexity Theory,* (1987), 160-173. (Also to appear in *Theor Comput Sci.*)

Related Papers - Other Authors

[Al-88] Allender, E., "Isomorphisms and *1-L* reductions," *J Comput System Sci* **36** (1988), 336-350. (Some of the results of this paper were first reported in *Proc 1st Symp Structure Complexity Theory, Springer Verlag LNCS,* **223** (1986), 12-22.

[Be-77] Berman, L., *Polynomial Reducibilities and Complete Sets,* PhD Thesis, Cornell Univ, (1977).

[Do-78] Dowd, M., "On isomorphism," Unpublished manuscript (1978).

[Do-82] Dowd, M., "Isomorphism of complete sets," *Rutgers U. (Busch Campus) Lab for CS Research Tech Report,* **34,** (1982).

[FKR-89] Fenner, S., S. Kurtz, and J. Royer, "Every polynomial-time 1-degree collapses iff $P = PSPACE$," *Proc 30th IEEE Symp Found Comput Sci,* (1989), 624-629.

[GH-90] Ganesan, K., "Complete problems, creative sets and isomorphism conjectures," *Ph.D. Dissertation,* Comput Sci Dept, Boston University, (1990), 1-76.

[GH-88] Ganesan, K. and S. Homer, "Complete problems and strong polynomial reducibilities," *Boston Univ Tech Report,* **88-001,** (1988).

[GJ-86] Goldsmith, J. and D. Joseph, "Three results on the polynomial isomorphism of complete sets," *Proc 27th IEEE FOCS,* (1986), 390-397.

[Go-86] Goldsmith, J., "Polynomial isomorphisms and near-testable sets," PhD Thesis, Univ Wisconsin, (1988); paper on collapsing degrees in preparation.

[GS-84] Grollmann, J. and A. Selman, "Complexity measures for public key cryptosystems," *Proc 25th IEEE FOCS,* (1984), 495-503.

[Ho-86] Homer, S., "On simple and creative sets in NP," *Theor Comput Sci* **47** (1986), 169-180.

[HS-89] Homer, S. and A. Selman, "Oracles for structural properties: the isomorphism problem and public-key cryptography," *Proc 4th IEEE Symp Structure Complexity Theory,* (1989), 3-14.

[JY-85] Joseph, D. and P. Young, "Some remarks on witness functions for nonpolynomial and noncomplete sets in NP," *Theor Comp Sci* **39** (1985), 225-237. The results in this paper were first reported in the survey [Yo-83].

[KLD-86] Ko, K., T. Long and D. Du, "A note on one-way functions and polynomial-time isomorphisms," *Theor Comput Sci* **47** (1986), 263-276. The results of this paper were first reported in *Proc 18th STOC,* 1986, 295-303.

[KM-81] Ko, K. and D. Moore, "Completeness, approximation and density," *SIAM J Comput* **10** (1981), 787-796.

[KMR-86] Kurtz, S., S. Mahaney and J. Royer, "Collapsing degrees," Extended abstract in *Proc 27th FOCS* (1986), 380-389. To appear in *J Comput Sys Sci.*

[KMR-87] Kurtz, S., S. Mahaney and J. Royer, "Non-collapsing degrees," *Univ Chicago Tech Report* **87-001.**

[KMR-87b] Kurtz, S., S. Mahaney and J. Royer, "Progress on collapsing degrees," *Proc 2nd IEEE Symp Structure in Complexity Theory,* (1987), 126-131.

[KMR-89] Kurtz, S., S. Mahaney and J. Royer, "The isomorphism conjecture fails relative to a random oracle," *Proc 21st ACM Symp Theory Comput* 1989, 157-166.

[KMR-90] Kurtz, S., S. Mahaney and J. Royer, "The structure of complete degrees," In A. Selman, editor, *Complexity Theory Retrospective,* pages 108–146, Springer Verlag, 1990.

[Lo-88] Long, T., "One-way functions, isomorphisms, and complete sets," invited presentation at Winter, 1988, AMS meetings (Atlanta), *Abstracts AMS,* Issue **55,** vol **9,** number 1, January 1988, 125.

[LS-86] Long, T. and A. Selman, "Relativizing complexity classes with sparse oracles," *JACM* **33** (1986), 618-627.

[Ma-81] Mahaney, S., "On the number of P-isomorphism classes of NP-complete sets," *Proc 22nd IEEE Symp Found Comput Sci,* (1981), 271-278.

[MY-85] Mahaney, S. and P. Young, "Reductions among polynomial isomorphism types," *Theor Comp Sci* **39** (1985), 207-224.

[My-55] Myhill, J., "Creative sets," *Z Math Logik Grundlagen Math* (1955), 97-108.

[Wa-89] Wang, J. "P-creative sets *vs.* P-completely creative sets," *Proc* **4**th *IEEE Symp Structure in Complexity Theory,* (1989), 24-35.

[Wa-85] Watanabe, O., "On one-one polynomial time equivalence relations," *Theor Comput Sci* **38** (1985), 157-165.

[Wa-87a] Watanabe, O., "On the structure of intractable complexity classes," PhD Dissertation, Tokyo Institute of Technology, 1987.

[Wa-87b] Watanabe, O., "Some observations of *k*-creative sets," Unpublished manuscript (1987).

[Yo-66] Young, P., "Linear orderings under one-one reducibility," *J Symbolic Logic* **31** (1966), 70-85.

[Yo-83] Young, P., "Some structural properties of polynomial reducibilities and sets in NP," *Proc* **15**th *ACM Symp Theory Computing,* (1983) 392-401. This is an overview of research which appears in journal form in [JY-85], [MY-85], and [LY-88].

3.5.4 Results on Sparse Sets and Kolmogorov Complexity Related to the Berman-Hartmanis Conjecture

Primary Papers - J. Hartmanis

[HM-80] Hartmanis, J. and S. Mahaney, "An essay about research on sparse NP-complete sets," *Proc* **9**th *Symp Math Found Computer Sci, Springer Verlag Lecture Notes Comput Sci* **88,** (1980), 40-57.

[Ha-83 b] Hartmanis, J., "Generalized Kolmogorov complexity and the structure of feasible computations," *Proc* **24**th *FOCS,* (1983), 439-445. Similar and related results also appear in, "On non-isomorphic NP-complete sets," *Bull EATCS* **24** (1984), 73-78.

Related Papers - J. Hartmanis

[Ha-63a] Hartmanis, J., "The equivalence of sequential machine models," *IEEE Trans Elect Comput* **EC-12,** (1963), 18-19.

[Ha-63b] Hartmanis, J., "Further results on the structure of sequential machines," *J Assoc Comput Mach* **10** (1963), 78-88.

[BH-79] Baker, T. and J. Hartmanis, "Succinctness, verifiability and determinism in representations of polynomial-time languages," *Proc* **20**th *IEEE Symp Found Comput Sci,* (1979), 392-396.

[Ha-80] Hartmanis, J., "On the succinctness of different representations of languages," *SIAM J Comput* **9** (1980), 114-120. This paper extends work in a paper by the same title presented at *ICALP* in 1979.

[Ha-81] Hartmanis, J., "Observations about the development of theoretical computer science," *Annals History Comput* **3** (1981), 42-51. The work in this paper was originally presented at the ^{20th} *FOCS,* (1979), in a paper with the same title.

[Ha-83] Hartmanis, J., "On Gödel speed-up and succinctness of language representations," *Theor Comput Sci* **26** (1983), 335-342.

[HIS-85] Hartmanis, J., N. Immerman and V. Sewelson, "Sparse sets in NP-P: EXPTIME versus NEXPTIME," *Infor and Control* 65 (1985), 159-181. The work in this paper was first presented under the same title at the **15**th *ACM Symp Theory Comput* in 1983.

[Ha-87a] Hartmanis, J., "Sparse complete sets for NP and the optimal collapses of the polynomial time hierarchy," *Bull EATCS* **32** - Structural Complexity Theory Column, (June 1987), 73-81.

[Ha-87b] Hartmanis, J., "The collapsing hierarchies," *Bull EATCS* **33** - Structural Complexity Theory Column, (October 1987), 26-39.

Related Papers - Other Authors

[Be-78] Berman, P., "Relationship between density and deterministic complexity of NP-complete languages," *Proc* **5**th *ICALP, Springer Verlag LNCS,* **62** (1978), 63-71.

[Bo-88] Book, R., "Sparse sets, tally sets, and polynomial reducibilities," invited talk, MFCS summer 1988, to appear:

[BWSD-77] Book, R., C. Wrathall, A. Selman and D. Dobkin, "Inclusion complete tally languages and the Berman-Hartmanis conjecture," *Math Systems Theory* **11** (1977), 1-8.

[Fo-79] Fortune, S., "A note on sparse complete sets," *SIAM J Comput* (1979), 431-433.

[Hr-88] Hromkovič, J., "Two independent solutions of the 23 years old open problem in one year, or *NSpace* is closed under complementation by two authors," *Bulletin EATCS* **34** (1988), 310-312.

[Im-88] Immerman, N., "Nondeterministic space is closed under complementation," *Proc 3rd IEEE Symp Structure Complexity Theory*, (1988), 112-115

[Ka-88] Kadin, J., "The polynomial hierarchy collapses if the boolean hierarchy collapses," *Proc 3rd IEEE Structure Complexity Theory*, (1988), 278-292.

[KL-80] Karp, R. and R. Lipton, "Some connections between uniform and nonuniform complexity classes," *Proc. of the 12th ACM STOC*, 1980, 302-309. (Also appears as "Turing machines that take advice," *L'Enseignement Mathématique* **82** (1982), 191-210.)

[Ko-83] Ko, K., "On the notion of infinite pseudorandom sequences," *Theor Comput Sci* **48** (1986), 9-33. The work presented in this paper first appeared in 1983 in an unpublished manuscript entitled, "Resource-bounded program-size complexity and pseudo-random sequences," and is often referenced in that form.

[Le-73] Levin, L., "Universal sequential search problems," *Prob Info Transmission* **9** (1973), 265-266.

[Le-84] Levin, L., "Randomness conservation inequalities; information and independence mathematical theories," *Inform and Control* **61** (1984), 15-37.

[LV-90] Li M. and P. Vitányi, "Applications of Kolmogorov Complexity in the Theory of Computation," A. Selman, editor, "Complexity Theory Retrospective," Springer Verlag, 1990, this volume. This paper expands on the article "Two decades of applied Kolmogorov complexity," in *Proc 3rd IEEE Symp Struct Complexity Theory*, (1988), 80-101.

[Lo-82] Long, T., "A note on sparse oracles for NP," *J Comput Systems Sci* **24** (1982), 224-232.

[Lo-86] Longpré, L., "Resource bounded Kolmogorov complexity, a link between computational complexity and information theory," *Cornell University Ph.D. Thesis,* Tech Report **TR-86-776,** (1986).

[LY-88] Longpré, L., and P. Young, "Cook is faster than Karp: A study of reducibilities in NP," *Proc* **3**rd *IEEE Symp Structure Complexity Theory,* (1988), 293-302. Also, *JCSS,* to appear.

[Ma-82] Mahaney, S., "Sparse complete sets for NP: solution of a conjecture of Berman and Hartmanis," *J Comput Systems Sci* **25** (1982), 130-143. The work in this paper first appeared in a paper of the same title presented at *FOCS* in 1980.

[Ma-89] Mahaney, S., "The isomorphism conjecture and sparse sets," in *Computational Complexity Theory,* edited by J. Hartmanis, *AMS Proc Symp Applied Math Series,* (1989). (This paper is an update and revision of "Sparse sets and reducibilities," in *Studies in Complexity Theory,* edited by R. Book, (1986), 63-118.)

[SW-88] Schöning, U. and K. Wagner, "Collapsing oracle hierarchies, census functions and logarithmically many queries," *Proc STACS,* (1988), *Springer Verlag LNCS,* **294**, 91-97.

[Si-83] Sipser, M., "A complexity theoretic approach to randomness" *Proc* **15**th *ACM Symp Theory of Comput* (1983), 330-335.

[Si-83] Szelepcsényi, R., "The method of forced enumeration for nondeterministic automata," *ACTA Informatica* **26** (1988), 279-284.

4

Describing Graphs: A First-Order Approach to Graph Canonization

Neil Immerman[1]
Eric Lander[2]

ABSTRACT In this paper we ask the question, "What must be added to first-order logic plus least-fixed point to obtain exactly the polynomial-time properties of unordered graphs?" We consider the languages $\mathcal{L}_k$ consisting of first-order logic restricted to k variables and $\mathcal{C}_k$ consisting of $\mathcal{L}_k$ plus "counting quantifiers". We give efficient canonization algorithms for graphs characterized by $\mathcal{C}_k$ or $\mathcal{L}_k$. It follows from known results that all trees and almost all graphs are characterized by $\mathcal{C}_2$.

4.1 Introduction

In this paper we present a new and different approach to the graph canonization and isomorphism problems. Our approach involves a combination of complexity theory with mathematical logic. We consider first-order languages for describing graphs. We define what it means for a language to characterize a set of graphs (Definition 4.4.2). We next define the languages $\mathcal{L}_k$ (resp. $\mathcal{C}_k$) consisting of the formulas of first-order logic in which only k variables occur (resp. $\mathcal{L}_k$ plus 'counting quantifiers'). We then study which sets of graphs are characterized by certain $\mathcal{L}_k$'s and $\mathcal{C}_k$'s. It follows by a result of Babai and Kučera [4] that the language $\mathcal{C}_2$ characterizes almost all graphs. We also show that $\mathcal{C}_2$ characterizes all trees. In Section 4.9 we give a simple $O[n^k \log n]$ step algorithm to test if two graphs G and H on n vertices agree on all sentences in $\mathcal{L}_k$, or $\mathcal{C}_k$. If G is characterized by $\mathcal{L}_k$ (or $\mathcal{C}_k$), a variant of this algorithm computes a canonical labeling for G in

[1]Computer and Information Science, University of Massachusetts, Amherst, MA 01003. Research supported by NSF grants DCR-8603346 and CCR-8806308. Part of this work was done in the Fall of 1985 while this author was visiting the Mathematical Sciences Research Institute, Berkeley, CA.

[2]Whitehead Institute, Cambridge, MA 02142 and Harvard University, Cambridge, MA 02138. Research supported by grants from the National Science Foundation (DCB-8611317) and from the System Development Foundation (G612).

the same time bound.

This line of research has two main goals. First, finding a language appropriate for graph canonization is a basic problem, central to the first author's work on descriptive computational complexity. We will explain this setting in the next section.

A canonization algorithm for a set of graphs, S, gives a unique ordering (canonical labeling) to each isomorphism class from S. Thus two graphs from S are isomorphic if and only if they are identical in the canonical ordering. The second goal of this work is to describe a simple and general class of canonization algorithms. We hope that variants of these algorithms will be powerful enough to provide simple canonical forms for all graphs; and do so without resorting to the the high powered group theory needed in the present, best graph isomorphism algorithms [27,3].

4.2   Descriptive Complexity

In this section we discuss an alternate view of complexity in which the complexity of the descriptions of problems is measured. This approach has provided new insights and techniques to help us understand the standard complexity notions: time, memory space, parallel time, number of processors. The motivations for the present paper come from Descriptive Complexity. We can only sketch this area here. The interested reader should consult [24] for a more extensive survey.

Given a property, S, one can discuss the computational complexity of checking whether or not an input satisfies S. One can also ask, "What is the complexity of *expressing* the property S?" It is natural that these two questions are related. However, it is startling how closely tied they are when the second question refers to expressing the property in first-order logic. We will now describe the first-order languages in detail. Next we will state some facts relating descriptive and computational complexity.

First-Order Logic

In this paper we will confine our attention to graphs and properties of graphs. Thus when we mention complexity classes P, NP, etc. we will really be referring to those problems of ordered graphs that are contained in P, NP, etc. (If you want to know why the word "ordered" was included in the previous sentence, please read on. One of the main concerns of this paper is how to remove the need to order the vertices of graphs.)

For our purposes, a *graph* will be defined as a finite logical structure, $G = \langle V, E \rangle$. V is the universe (the vertices); and E is a binary relation on V (the edges). As an example, the undirected graph, $G_1 = \langle V_1, E_1 \rangle$, pictured in Figure 1 has vertex set $V_1 = \{0, 1, 2, 3, 4\}$, and edge relation $E_1 = \{\langle 0, 3 \rangle, \langle 0, 4 \rangle, \langle 1, 2 \rangle, \langle 1, 3 \rangle, \dots, \langle 3, 2 \rangle, \langle 3, 4 \rangle\}$ consisting of 12 pairs cor-

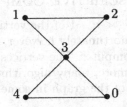

FIGURE 4.1. An Undirected Graph

responding to the six undirected edges. By convention, we will assume that all structures referred to in this paper have universe, $\{0, 1, \ldots, n-1\}$ for some natural number n.

The *first-order language* of graph theory is built up in the usual way from the variables, $x_1, x_2, \ldots$, the relations symbols, E and $=$, the logical connectives, $\wedge, \vee, \neg, \rightarrow$, and the quantifiers, $\forall$ and $\exists$. The quantifiers range over the vertices of the graph in question. For example consider the following first-order sentence:

$$\varphi \equiv \forall x \forall y \left[E(x, y) \rightarrow E(y, x) \wedge x \neq y \right]$$

φ says that G is undirected and loop free. Unless we specifically say otherwise, we will assume that all graphs, G, in this paper satisfy φ, in symbols: $G \models \varphi$.

It is useful to consider an apparently[3] more general set of structures. The *first-order language of colored graphs* consists of the addition of a countable set of unary relations $\{C_1, C_2, \ldots\}$ to the first-order language of graphs. Define a *colored graph* to be a graph that interprets these new unary relations so that all but finitely many of the predicates are false at each vertex. These unary relations may be thought of as colorings of the vertices. (A vertex of a colored graph may satisfy zero, one, or several of the color relations. However, we will say that two vertices are the same color iff they satisfy the same *set* of color relations. Thus, by increasing the number of color relations we may assume that each vertex satisfies a unique color relation.)

---

[3]Colorings can be simulated in uncolored graphs by attaching gadgets. For example, a colored graph G with colors green and yellow can be modeled as a graph G' with some auxiliary vertices so that in G' each vertex v from G is now connected to either a triangle, or a square, or a pentagon, or a hexagon, according as v is green, yellow, green and yellow, neither green nor yellow. All mention of color predicates in this paper can be removed in this way.

RESULTS FROM DESCRIPTIVE COMPLEXITY

The ordering $0 < 1 < \cdots < n-1$ of the vertices of a graph is irrelevant to a "graph property". Unfortunately however, it is impossible to present an unordered graph to a computer. The vertices and edges must be presented in some order. Furthermore, many algorithms compute graph properties by visiting the vertices of the graph in some order which depends on this arbitrary ordering.

Let FO($\leq$) denote the first-order language of ordered graphs. This language includes the logical relation $\leq$ which must be interpreted as the usual ordering on the vertices, $V = \{0, 1, \ldots, n-1\}$. We will see in Fact 4.2.8 that FO($\leq$) is contained in CRAM[1] – the set of properties that can be checked by a concurrent, parallel, random access machine in constant time, using polynomially many processors.

In order to express a richer class of problems we consider uniform[4] sequences of first-order sentences, $\{\varphi_i\}_{i \in \mathbf{N}}$, where the sentence φ_n expresses the property in question for graphs with at most n vertices.

Let FO($\leq$)$[t(n)] - \mathrm{VAR}[v(n)]$ be the set of problems expressible by a uniform sequence of first-order sentences such that the nth sentence has length $O[t(n)]$ and uses at most $v(n)$ distinct variables. The following fact says that the set of polynomial-time recognizable properties of ordered graphs is equal to the set of properties expressible by uniform sequences of first-order sentences of polynomial length using a bounded number of distinct variables.

Fact 4.2.1 ([18])

$$P = \bigcup_{k=1}^{\infty} \mathrm{FO}(\leq)[n^k] - \mathrm{VAR}[k]$$

The Least Fixed Point (LFP) operator has long been used by logicians to formalize the power to define new relations by induction, cf. [28]. In [19] and in [31] it is shown that the uniform sequence of formulas in Fact 4.2.1 can be represented by a single use of LFP applied to a first-order formula. Thus,

Fact 4.2.2

$$P = \mathrm{FO}(\leq, \mathrm{LFP}) = \bigcup_{k=1}^{\infty} \mathrm{FO}(\leq)[n^k] - \mathrm{VAR}[k]$$

---

[4]The uniformity in question can be purely syntactical, i.e. the nth sentences of a FO$[t(n)]$ property consists of a fixed block of quantifiers repeated $t(n)$ times followed by a fixed quantifier-free formula. Uniformity is discussed extensively in [6]. In this paper the reader may think of uniform as meaning that the map from n to φ_n is easily computable, e.g. in logspace.

Example 4.2.3 The *monotone circuit value problem* is an example of a complete problem for P which we will use to illustrate Fact 4.2.2, [13]. Instances of this problem consist of boolean circuits with only "and" and "or" gates and a unique output gate whose value as determined by the inputs is one.

We can code the monotone circuit value problem as a colored, directed graph with colors T, A, R as follows: inputs are colored *True* if they are on; other vertices are colored *And* if they are "and" gates; otherwise they are "or" gates. The unique output gate is colored *Root*. The obvious inductive definition of circuit value can be written in (FO + LFP) as follows:

$$\psi(V, x) \;\equiv\; T(x) \vee \Big[(\exists y)(E(x, y) \wedge V(y))$$
$$\wedge \, (A(x) \rightarrow (\forall y)(E(x, y) \rightarrow V(y)))\Big]$$

The intuitive meaning of $V(x)$ is that the value of node x is one. Thus $V(x)$ is true if x is an input that is on, or if x is an "or" gate and there exists a gate y such that $V(y)$ and y is an input to x, or if x is an "and" gate having at least one input, and all of its inputs, y, satisfy $V(y)$.

Fix a graph, C. The formula $\psi(V, x)$ induces an operator ψ_C from unary relations on the vertices of C to unary relations on the vertices of C as follows:

$$\psi_C(R) \;=\; \{z \mid C \models \psi(R, z)\}$$

Furthermore, the correct Value relation on the circuit C is the least fixed point of ψ_C, i.e. the smallest unary relation V such that $\psi_C(V) = V$. Use the notation (LFP $\psi(V, x)$) to denote this least fixed point.

Let $(\exists! x)\alpha$ (there exists a unique x such that α) be an abbreviation for the formula,

$$(\exists x)(\alpha(x) \wedge (\forall y)(\alpha(y) \rightarrow y = x))$$

The circuit value problem can now be expressed as follows:

$$(\exists! w)(R(w)) \wedge (\exists w)(R(w) \wedge (\text{LFP } \psi(V, x))(w))$$

A particular kind of inductive operator that is worth studying on its own is the transitive closure operator (TC) introduced in [20]. Let $\varphi(\bar{x}, \bar{y})$ be any binary relation on k-tuples. Then $(\text{TC } \varphi(\bar{x}, \bar{y}))$ denotes the reflexive, transitive closure of φ. The following was proved in [20] with the finishing touch proved in [21].

Fact 4.2.4

$$\text{FO}(\leq, \text{TC}) = \text{NSPACE}[\log n]$$

Example 4.2.5 Consider the following complete problem for NSPACE[$\log n$]. The GAP problem consists of the set of directed graphs having a unique vertex colored S and a unique vertex colored T and such that there is a path from the S vertex to the T vertex. It is obvious how to express GAP in (FO + TC):

$$(\exists!x)(S(x)) \wedge (\exists!x)(T(x)) \wedge (\exists st)(S(s) \wedge T(t) \wedge (\text{TC } E(x,y))(x,t))$$

It is interesting to examine the relationship between the number of variables needed to describe a problem, and the computational complexity of the problem. Let $\text{FO}[t(n)] - \text{VAR}[v]$ be the restriction of $\text{FO}[t(n)]$ to sentences with at most v distinct variables. Then the following bounds can be derived from the proof of Fact 4.2.1 in [18]:

Fact 4.2.6 *[18]*

$$DTIME[n^k] \subseteq \text{FO}(\le)[n^k] - \text{VAR}[k+3] \subseteq DTIME[n^{2k+4}]$$

Thus the $DTIME[n^k]$ properties of ordered graphs are *roughly* the properties expressible by first-order sentences with k variables and length n^k. Obviously this is very rough. A closer relationship between machine complexity and first-order expressibility is obtained if one takes into account the built in parallelism of quantifiers.

Let $\text{CRAM}[t(n)]\text{-PROC}[p(n)]$ be the set of problems accepted by a concurrent-read, concurrent-write, parallel random access machine (CRAM) in parallel time $O[t(n)]$ using $O[p(n)]$ processors. In order to get a precise relationship with the CRAM model when $t(n) < \log n$ it was necessary to add another logical relation to FO. Since variables range over the universe $\{0, 1, \ldots, n-1\}$ they may be thought of as $\log n$ bit numbers. Let the relations $\text{BIT}(x,y)$ be true just if the xth bit in the binary expansion of y is a one. In [22] it is shown that $\text{FO}(\le, \text{BIT})[t(n)] - \text{VAR}[O[1]]$ is exactly the set of properties checkable by a CRAM in parallel time $O[t(n)]$ using polynomially many processors. (In fact, $\text{FO}(\le, \text{BIT})[t(n)] - \text{VAR}[v]$ corresponds to $\text{CRAM-TIME}[t(n)]$ using *roughly* n^v processors.)

Fact 4.2.7 *For all $t(n)$,*

$$\text{FO}(\le, \text{BIT})[t(n)] - \text{VAR}[O[1]] = \text{CRAM}[t(n)]\text{-PROC}[n^{O[1]}]$$

In particular, we have that the first-order properties are those checkable in constant time by a CRAM using polynomially many processors,

Fact 4.2.8

$$\text{FO}(\le, \text{BIT}) = \text{CRAM}[1]\text{-PROC}[n^{O[1]}]$$

4.3   Properties of (Unordered) Graphs

Facts 4.2.2 and 4.2.4 give natural languages expressing respectively the polynomial-time and nondeterministic logspace properties of ordered graphs.

When the ordering is not present, it is possible to prove nearly optimal upper and lower bounds on the number of quantifiers and variables needed to express various properties in first-order logic.

For example, in [18] the graphs Y_k and N_k are constructed. These graphs have the property that Y_k has a complete subgraph on k vertices, but N_k does not. However using Ehrenfeucht-Fraisse games (cf. Section 4.6) one can show that Y_k and N_k agree on all sentences with $k-1$ variables but without ordering. It thus follows that k variables are necessary and sufficient to express the existence of a complete subgraph of size k. If these bounds applied to the languages with ordering they would imply that $P \neq NP$.

In [17] there is a similar construction of a sequence of pairs of graphs which differ on a polynomial-time complete property, but agree on all sentences of poly-logarithmic length without ordering. If this result went through with the ordering it would follow that $NC \neq P$, and in particular that $NSPACE[\log n]$ is not equal to P.

The reason these arguments do not go through with ordering is as follows. For any constant c, there is a very simple formula[5] with ordering, $\alpha_c(x)$, that holds just for x equal to the cth vertex. It follows that whenever two graphs agree on all simple sentences with ordering, they are equal.

It is of great interest to understand the role of ordering and if possible to replace the ordering with a more benign construction. Furthermore, the most basic problem on which to study the role of ordering is graph isomorphism. If two graphs differ on any property they are certainly not isomorphic!.

Let a *graph property* be an order independent property of ordered graphs. One can ask the question,

Question 4.3.1 *Is there a natural language for the polynomial-time graph properties?*

Gurevich has conjectured that the answer to Question 4.3.1 is, "No," [14]. An affirmative answer to this question would imply a similar answer to the more basic,

Question 4.3.2 *Is there a recursively enumerable listing of all polynomial-time graph properties?*

Questions 4.3.1 and 4.3.2 are important in various settings. It is well known that graphs are the most general logical structures.[6] Thus these questions are equivalent to the corresponding questions concerning relational databases: i.e. give a database query language for expressing exactly

---

[5]More precisely, the formula has 3 variables and length $O[\log n]$.

[6]More precisely, every first-order language may be interpreted in the first-order theory of graphs. We would like to know to whom this is due, and where it appears in print.

the polynomial-time queries that are independent of the arbitrary ordering of tuples, cf. [9]. We believe that the answers to Questions 4.3.1 and 4.3.2 are both, "Yes," and we ask the more practical,

Question 4.3.3 *What must we add to first-order logic after taking out the ordering so that Fact 4.2.2 remains true? Put another way, describe a language $\mathcal{L}$ that expresses exactly the polynomial-time graph properties.*

The ordering relation is crucial for simulating computation: a Turing machine will be given an input graph in some order. It will visit the vertices of the graph using this ordering; and it is difficult to see how to simulate an arbitrary computation without reference to this ordering. It is well known that first-order logic without ordering is *not* strong enough to express computation. Let EVEN be the set of graphs with an even number of vertices. We will show in Proposition 4.6.4 that the property EVEN requires n variables for graphs with n vertices. (For a property to be expressible in FO+LFP a necessary condition is that it is expressible in a constant number of variables independent of n.)

In view of Proposition 4.6.4, it is natural to add the ability to count to first-order logic without ordering. This is formalized in Section 4.7, where we define the languages $\mathcal{C}_k$ of first-order logic restricted to k distinct variables, plus "counting quantifiers". We show in Corollary 4.8.5 that the very simple language $\mathcal{C}_2$ suffices to give unique descriptions and thus efficient canonical forms for almost all graphs.

For a long time we suspected that first-order logic plus least fixed point and counting was enough to express all polynomial-time graph properties. It would have immediately followed that for each polynomial-time graph property Q there would be a fixed k such that for all n, the property Q restricted to graphs of size n is expressible in $\mathcal{C}_k$. In particular, if our suspicion were right, then for every set of graphs S admitting a polynomial-time graph isomorphism algorithm, there would exist a fixed k such that $\mathcal{C}_k$ characterizes S (to be defined later). This implies that for any two graphs G and H from S, if G and H are $\mathcal{C}_k$ equivalent (i.e. G and H agree on all sentences from $\mathcal{C}_k$) then they are isomorphic. For example, the sets of graphs of bounded color class size (defined below) admit polynomial-time graph isomorphism algorithms. We show in Proposition 4.5.3 that the language $\mathcal{C}_3$ characterizes graphs of color class size 3. However, the following recent result shows in a strong way that no $\mathcal{C}_k$ characterizes the graphs of color class size 4. Thus our suspicion was wrong: first-order logic plus least fixed point and counting does not express all the polynomial-time graph properties.

Fact 4.3.4 ([8]) *There exists a sequence of pairs of non-isomorphic graphs $\{G_n, H_n\}$ such that G_n and H_n have $O[n]$ vertices, color class size 4, and admit linear time and logspace canonization algorithms. However, G_n and H_n are $\mathcal{C}_n$ equivalent.*

4.4   Characterization of Graphs

Throughout this paper we will be considering various languages for describing colored graphs. We are interested in knowing when a language suffices to characterize a particular graph, or class of graphs. Some of the following definitions and notation are adapted from [25].

Definition 4.4.1 For a given language $\mathcal{L}$ we say that the graphs G and H are $\mathcal{L}$-equivalent ($G \equiv_{\mathcal{L}} H$) iff for all sentences $\varphi \in \mathcal{L}$,

$$G \models \varphi \ \Leftrightarrow \ H \models \varphi.$$

A *partial valuation* over a graph $G = (V, E)$ is a partial function $u : \{x_1 \ldots\} \to V$. The domain of u is denoted ∂u. Call a $(k$-$)$*configuration* over G, H a pair (u, v) where u is a partial valuation over G and v is a partial valuation over H such that $\partial u = \partial v (\subseteq \{x_1, \ldots x_k\})$. If (u, v) is a k-configuration over G and H, we say that G, u and H, v are $\mathcal{L}$-*equivalent* ($G, u \equiv_{\mathcal{L}} H, v$) iff for all formula $\varphi \in \mathcal{L}$, with free variables from $x_1, \ldots, x_k$,

$$G, u \models \varphi \ \Leftrightarrow \ H, u \models \varphi.$$

Using the concept of $\mathcal{L}$-equivalence, we can now define what it means for the language $\mathcal{L}$ to characterize a set of graphs.

Definition 4.4.2 We say that $\mathcal{L}$ k-*characterizes* G iff for all graphs H, and for all k-configurations (u, v) over G, H, if G, u and H, v are $\mathcal{L}$-equivalent then there is an isomorphism from G to H extending the correspondence given by (u, v). In symbols,

$$(G, u \equiv_{\mathcal{L}} H, v) \ \Rightarrow \ (\exists f \supset v \circ u^{-1})(f : G \overset{\cong}{\to} H).$$

We say that $\mathcal{L}$ *characterizes* G iff $\mathcal{L}$ 1-characterizes $\hat{G}$, for all colorings $\hat{G}$ of G. For a set of graphs S, we say that $\mathcal{L}$ *characterizes* S iff for all $G \in S$, $\mathcal{L}$ characterizes G.

Proposition 4.4.3 *Let GRAPHS be the set of all finite, colored graphs, and let FO be the first-order language of colored graphs. Then FO characterizes GRAPHS.*

Proof   Let $G \in$ GRAPHS have n vertices, and let u be a partial valuation over G. For simplicity, suppose that $\partial u = \{x_1, \ldots, x_k\}$, and $u(x_1), \ldots, u(x_k)$ are all distinct. Let $g_1, \ldots, g_n$ be a numbering of G's vertices so that $g_i = u(x_i)$, for $1 \leq i \leq k$. Let r be a subscript greater than that of any color relation holding in G. It is simple to write a first-order formula, χ_r, with $n + 1 - k$ quantifiers that says (a) there exist $x_{k+1} \ldots x_n$ such that the x_i's are all distinct; (b) any other vertex is equal to one of the x_i's; (c) each pair (x_i, x_j) has an edge or not exactly as the edge (g_i, g_j) occurs or not in G; and finally (d) for each x_i, $i \leq n$, and each C_j, $j \leq r$, $C_j(x_i)$ holds exactly if $C_j(g_i)$ holds in G. Let H be any graph, and let r be greater than the index of any color relation holding in H. Let v be any valuation over H such that H, v satisfies χ_r. Let v' be an extension of v to a valuation over H with $\partial v' = \{x_1 \ldots x_n\}$, making the quantifier-free part of χ_r true. Then clearly $f : g_i \mapsto v'(x_i)$ is the desired isomorphism. ∎

Proposition 4.4.3 leads to an inefficient graph canonization algorithm. In the next section, we consider languages weaker than full first-order logic, in order to obtain efficient algorithms.

4.5   The Language $\mathcal{L}_k$

Define $\mathcal{L}_k$ to be the set of first-order formulas, φ, such that the quantified variables in φ are a subset of $x_1, x_2, \ldots, x_k$. Note that variables in first-order formulas are similar to variables in programs: they can be reused (i.e. requantified). For example consider the following sentence in $\mathcal{L}_2$.

$$\psi \;\; \equiv \;\; \forall x_1 \exists x_2 \Big(E(x_1, x_2) \wedge \exists x_1 \big[\neg E(x_1, x_2) \big] \Big)$$

The sentence, ψ, says that every vertex is adjacent to some vertex which is itself not adjacent to every vertex. As an example, the graph from Figure 1 satisfies ψ. Note that the outermost quantifier, $\forall x_1$, refers only to the free occurrence of x_1 within its scope.

In this section we will consider the question, "Which graphs are characterized by $\mathcal{L}_k$?" Define a *color class* to be the set of vertices which satisfy a particular set of color relations and no others. The *color class size* of a graph is defined to be the cardinality of the largest color class.

Proposition 4.5.1 $\mathcal{L}_2$ *characterizes the colored graphs with color class size one.*

Proof   This is clear. In $\mathcal{L}_2$ we can assert that each color class is of size at most one, e.g. $\forall x_1 \forall x_2 \big(B(x_1) \wedge B(x_2) \rightarrow x_1 = x_2 \big)$. We can also say which edges exist, e.g. the blue vertex is connected to the red vertex. Thus if graph G has color class size one, and if $G, g \equiv_{\mathcal{L}_2} H, h$ then there is an isomorphism $f : G \to H$. Since f preserves colors, $f(g) = h$. ∎

Next we consider the much more powerful language $\mathcal{L}_3$. In this language we can express the existence of paths.

Proposition 4.5.2 *For any natural number r, the formula $P_r(x_1, x_2)$, meaning that there is a path of length at most r from x_1 to x_2, can be written in $\mathcal{L}_3$.*

Proof By induction. $P_1(x_1, x_2)$ is $E(x_1, x_2) \vee x_1 = x_2$. Inductively,

$$P_{s+t}(x_1, x_2) \equiv \exists x_3 \big(P_s(x_1, x_3) \wedge P_t(x_3, x_2) \big)$$

Note that a maximum of 3 distinct variables is used. ∎

We will see in Section 4.6 that there are graphs with color class size 2 that cannot be distinguished by a sentence in $\mathcal{L}_2$. The ability of $\mathcal{L}_3$ to talk about path lengths makes it slightly less trivial:

Proposition 4.5.3 $\mathcal{L}_3$ *characterizes graphs of color class size at most three.*

Proof Let G and H be colored graphs, let g and h be vertices of G and H, and suppose that $G, g \equiv_{\mathcal{L}_3} H, h$. We will build an isomorphism $f : G \to H$, such that $f(g) = h$.

We first refine the colorings of the vertices of G and H to correspond to $\mathcal{L}_3$ types. For $A, B \in \{G, H\}$, vertices $a \in A$ and $b \in B$ will have the same $\mathcal{L}_k$-refined color iff they satisfy the same $\mathcal{L}_k$ formulas, i.e.

$$\{ \varphi \in \mathcal{L}_k \mid A \models \varphi_a^{x_1} \} = \{ \varphi \in \mathcal{L}_k \mid B \models \varphi_b^{x_1} \}^7 .$$

The following lemma says that we may assume that the color types of G and H are already refined.

Lemma 4.5.4 *Let the finite, colored graphs G and H be $\mathcal{L}_k$ equivalent and let G' and H' be the $\mathcal{L}_k$ color refinements of G and H. Then G' and H' are $\mathcal{L}_k$ equivalent.*

Proof Since G and H are finite, each refined color class C_i' is determined by the conjunction $\psi_i \in \mathcal{L}_k$ of a finite set of formulas. That is for all i, G' and H' both satisfy

$$\forall x_1 (C_i'(x_1) \leftrightarrow \psi_i) .$$

Note that ψ_i has x_1 as its free variable. Thus any occurrence of $C_i'(x_1)$ may be replaced by the equivalent ψ_i. Similarly any occurrence of $C_i'(x_j)$, $j = 2, \ldots, k$ may be replaced by $\psi_i^{\pi_j}$ where π_j is a permutation of $\{x_1, \ldots, x_k\}$ sending x_1 to x_j. Now for any formula $\alpha \in \mathcal{L}_3(C_1', C_2', \ldots)$ we may replace each occurrence of $C_i'(x_j)$ by $\psi_i^{\pi_j}$ to obtain an equivalent formula $\alpha' \in \mathcal{L}_3(C_1, \ldots C_r)$. ∎

By the above lemma we may assume that the color classes of G and H correspond exactly to the $\mathcal{L}_3$ types of the vertices. Let R and B be two colors and consider the edges between red and blue vertices in G or H. Note

---

[7]The notation α_t^x denotes the formula α with the term t substituted for the variable x.

that this is a regular bipartite graph because we can express in $\mathcal{L}_3$ that a red vertex has $0, 1, 2$, or all blue vertices as neighbors. Note also that for color classes of size at most 3, the only regular bipartite graphs representing nontrivial relationships between vertices are the 1:1 correspondence graphs and their complements. Let us then change such bipartite graphs as follows: replace the complete bipartite graph by its complement, and replace the graphs of degree two whose complements are 1:1 correspondence graphs by these complements. Note that when we perform these changes on G and H the new graphs are still $\mathcal{L}_3$ equivalent, and they are isomorphic now iff they were before.

Let the *color valence* of a graph be the maximum number of edges from any vertex to vertices of a fixed color. We have reduced the problem to constructing an isomorphism between $\mathcal{L}_3$-equivalent graphs G and H when these graphs have color valence one. We construct the isomorphism f as follows: Begin by letting $f(g) = h$. Next, while there is a vertex g_1 in the domain of f with a (unique) neighbor g_2 of color C_i not yet in the domain of f, do the following. Let h_2 be the neighbor of $f(g_1)$ of color C_i, and let $f(g_2) = h_2$.

We claim that the function f constructed above is an isomorphism from G to H. If not, then it must be the case that there is a loop of a certain color sequence in one of the graphs but not the other. For example, suppose that we chose $g_1, g_2, \ldots, g_j$ and $h_1, h_2, \ldots, h_j$ so that g_1 and h_1 are color C_1, and for $i < j$, g_{i+1} and h_{i+1} are the unique neighbors of g_i and h_i, respectively, of color C_{i+1}. However, suppose now that the neighbor of h_j of color C_1 is h_1, but that g_1 is not a neighbor of g_j. In this case there is a certain easily describable loop in H but not in G. That means that G and H disagree on the following $\mathcal{L}_3$ formula:

$$\Big(C_1(x_1) \quad \wedge \quad \exists x_2\big(C_2(x_2) \wedge E(x_1, x_2) \wedge \exists x_3(C_3(x_3) \wedge E(x_2, x_3) \wedge$$

$$\wedge \quad \exists x_2(C_4(x_2) \wedge E(x_3, x_2) \wedge \ldots \wedge \exists x_i(C_j(x_i) \wedge E(x_i, x_1))\ldots)\Big)$$

Since $G \equiv_{\mathcal{L}_3} H$ they must agree on the above formula. Therefore f is an isomorphism as claimed.   ∎

In the next section we describe some games that may be used to prove lower bounds on the expressibility of the $\mathcal{L}_k$'s. We will show as an example using these games that $\mathcal{L}_2$ does not suffice to characterize graphs of color class size 2. Recently it has been shown (cf. Fact 4.3.4) that no fixed $\mathcal{L}_k$ suffices to characterize the graphs of color class size 4.

4.6   Lower Bounds

In this section we will show that $\mathcal{L}_k$ is not expressive enough to characterize graphs efficiently. We will use the combinatorial games of Ehrenfeucht and

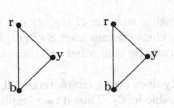

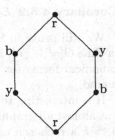

FIGURE 4.2. The $\mathcal{L}_2$ Game

Fraisse [10,12] as modified for $\mathcal{L}_k$ (see [18,7,29]). All of the results in this section could be proved by induction on the complexity of the sentences in question; but, we find that the games offer more intuitive arguments.

Let G and H be two graphs, and let k be a natural number. Define the $\mathcal{L}_k$ game on G and H as follows. There are two players, and there are k pairs of pebbles, $g_1, h_1, \ldots, g_k, h_k$. On each move, Player I picks up any of the pebbles and he places it on a vertex of one of the graphs. (Say he picks up g_i. He must then place it on a vertex from G.) Player II then picks up the corresponding pebble, (If Player I chose g_i then she must choose h_i), and places it on a vertex of the appropriate graph (H in this case).

Let $p_i(r)$ be the vertex on which pebble p_i is sitting just after move r. Then we say *Player I wins the game at move r* if the map that takes $g_i(r)$ to $h_i(r)$, $i = 1, \ldots, k$, is not an isomorphism of the induced k vertex subgraphs. Note that if the graphs are colored then an isomorphism must preserve color as well as edges. Thus Player II has a winning strategy for the $\mathcal{L}_k$ game just if she can always find matching points to preserve the isomorphism. Player I is trying to point out a difference between the two graphs and Player II is trying to keep them looking the same.

As an example consider the $\mathcal{L}_2$ game on the graphs G and H shown in Figure 2.

Suppose that Player I's first move is to place g_1 on a red vertex in G. Player II may answer by putting h_1 on either of the red vertices in H. Now suppose Player I puts h_2 on an adjacent yellow vertex in H. Player II has a response because in G, $g_1(1)$ also has an adjacent yellow vertex. The reader should convince himself or herself that in fact Player II has a winning strategy for the $\mathcal{L}_2$ game on the given G and H. The relevant theorem concerning the relationship between this game and the matter at hand is:

Fact 4.6.1 *[18, Theorem C.1] Let (u, v) be a k-configuration over G, H. Player II has a winning strategy for the $\mathcal{L}_k$ game on (u, v) if and only if $G, u \equiv_{\mathcal{L}_k} H, v$.*

Note that we have the following

Corollary 4.6.2 $\mathcal{L}_2$ *does not characterize graphs of color class size 2.*

We will prove in Section 4.9 that testing whether $G \equiv_{\mathcal{L}_k} H$ can be done in time $O[n^k \log n]$. Furthermore, if $\mathcal{L}_k$ characterizes a set S of graphs, then canonical forms for the graphs in S may be computed in this same time bound.

It is interesting to note that not only does no $\mathcal{L}_k$ characterize all graphs, but almost all graphs are indistinguishable in $\mathcal{L}_k$. Thus if two graphs of size $n \gg k$ are chosen at random they will almost certainly be $\mathcal{L}_k$ equivalent, but not isomorphic.

Fact 4.6.3 *[18],cf [11] Fix k and let $Pr_n(G \equiv_{\mathcal{L}_k} H)$ be the probability that two randomly chosen graphs of size n are $\mathcal{L}_k$ equivalent. Then*

$$\lim_{n \to \infty} \Big[Pr_n(G \equiv_{\mathcal{L}_k} H) \Big] = 1$$

Not only does $\mathcal{L}_k$ not characterize most graphs, it is not strong enough to express counting:

Proposition 4.6.4 *Let EVEN be the set of graphs with an even number of vertices. This property is not expressible in $\mathcal{L}_n$ for graphs with n or more vertices. Furthermore, $\mathcal{L}_n$ does not characterize the set of totally disconnected graph on n vertices.*

Proof Let D_n be the uncolored graph with n vertices and no edges. We claim that $D_n \equiv_{\mathcal{L}_n} D_{n+1}$. The following is a winning strategy for Player II in the n-pebble game on D_n and D_{n+1}. Player I's moves are answered preserving distinctness. That is, if Player I places pebble i on a vertex already occupied by pebble j, then Player II does the same. If Player I places pebble i on a vertex not occupied by any other pebbles, then Player II does the same. This is possible, because there are n vertices, and only $n-1$ other pebbles. Since there are no edges, the resulting maps are always isomorphisms. ∎

In the next section we increase the expressive power of the $\mathcal{L}_k$'s by adding the ability to count.

4.7   Counting Quantifiers

In this section we add *counting quantifiers* to the languages $\mathcal{L}_k$, thus obtaining the new languages $\mathcal{C}_k$. For each positive integer, i, we include the quantifier, $(\exists i\, x)$. The meaning of "$(\exists 17\, x_1)\varphi(x_1)$", for example, is that there exist at least 17 vertices such that φ. We will sometimes also use the quantifiers, $(\exists! i\, x)$, meaning that there exists exactly i x's:

$$(\exists! i\, x)\varphi(x) \quad \equiv \quad (\exists i\, x)\varphi(x) \;\wedge\; \neg(\exists i + 1\, x)\varphi(x)$$

Example 4.7.1 As our first example, note that the following sentence in $\mathcal{C}_2$ characterizes the graph D_n of Proposition 4.6.4:

$$(\exists! n\, x)(x = x) \;\wedge\; (\forall x)(\forall y)(\neg E(x,y)) \;.$$

Note that every sentence in $\mathcal{C}_k$ is equivalent to an ordinary first-order sentence with perhaps many more variables and quantifiers. We will see that testing $\mathcal{C}_k$ equivalence is no harder than testing $\mathcal{L}_k$ equivalence – the idea is that to test the truth of $\forall x$ or $\exists x$ we have to consider all possible x's anyway, and it doesn't cost more to count them. In Corollary 4.9.7 we show that $\mathcal{C}_k$ equivalence can be tested in time $O[n^k \log n]$. Similarly, graphs characterized by $\mathcal{C}_k$ can be given canonical labelings in the same time.

The following notation is useful.

Definition 4.7.2 Let Σ be a set of finite graphs. Define $\mathrm{var}(\Sigma, n)$ (resp. $\mathrm{vc}(\Sigma, n)$) to be the minimum k such that $\mathcal{L}_k$ (resp. $\mathcal{C}_k$) characterizes the graphs in Σ with at most n vertices. Let $\mathrm{var}(n) = \mathrm{var}(GRAPHS, n)$ and $\mathrm{vc}(n) = \mathrm{vc}(GRAPHS, n)$. When $\mathrm{var}(\Sigma, n)$ or $\mathrm{vc}(\Sigma, n)$ is bounded, we write $\mathrm{var}(\Sigma) = \max_n \mathrm{var}(\Sigma, n)$, and $\mathrm{vc}(\Sigma) = \max_n \mathrm{vc}(\Sigma, n)$.

For example, by combining various results obtained so far we know that $\mathrm{var}(GRAPHS, n) = n + 1$, $\mathrm{var}(CC1) = 2$, and $\mathrm{var}(CC2) = \mathrm{var}(CC3) = 3$. Here we are letting CCk be the set of color class k graphs.

We will now examine $\mathcal{C}_k$, attempting to compute $\mathrm{vc}(S)$ for various sets of graphs, S. A modification of the $\mathcal{L}_k$ game provides a combinatorial tool for analyzing the expressive power of $\mathcal{C}_k$. Given a pair of graphs define the $\mathcal{C}_k$ game on G and H as follows: Just like the $\mathcal{L}_k$ game we have two players and k pairs of pebbles. Now however each move has two steps.

1. Player I picks up a pebble (say g_i). He then chooses a set, A, of vertices from one of the graphs (in this case G). Now Player II must answer with a set, B, of vertices from the other graph. B must have the same cardinality as A.

2. Player I places h_i on some vertex $b \in B$. Player II answers by placing g_i on some $a \in A$.

The definition for winning is as before. Note that what is going on in the two step move is that Player I is asserting that there exist $|A|$ vertices in G with a certain property. Player II answers that there are the same number of such vertices in H. A straight forward extension of the proof of Fact 3.1 shows that this game does indeed capture expressibility in $\mathcal{C}_k$.

Theorem 4.7.3 *Let (u, v) be a k-configuration over G, H. Player II has a winning strategy for the $\mathcal{C}_k$ game on (u, v) if and only if $G, u \equiv_{\mathcal{C}_k} H, v$.*

Consider the following example of the $\mathcal{C}_k$ game.

Proposition 4.7.4 *Player II has a win for the $\mathcal{C}_2$ game on the graphs pictured in Figure 2. Thus $vc(CC2) > 2$.*

Proof   Player II's winning strategy is as follows: She matches the first vertex chosen by Player I with any vertex of the same color. Now suppose that at any point in the game, the first pair of pebbles are placed on vertices g_1 and h_1, both vertices of the same color, say red. Suppose that Player I's next move involves the other pair of pebbles. There is a 1:1 correspondence between the vertices in G and H as follows:

$$g_1 \qquad\qquad\qquad\qquad \mapsto h_1$$

blue vertex adjacent to g_1 $\qquad \mapsto$ blue vertex adjacent to h_1
yellow vertex adjacent to g_1 $\qquad \mapsto$ yellow vertex adjacent to h_1
red vertex not adjacent to g_1 $\qquad \mapsto$ red vertex not adjacent to h_1
yellow vertex not adjacent to $g_1 \mapsto$ yellow vertex not adjacent to h_1
blue vertex not adjacent to g_1 $\quad \mapsto$ blue vertex not adjacent to h_1

If Player I chooses a set A, then Player II chooses the set B to be the corresponding set of vertices under the above map. Whichever vertex Player I then picks from B, Player II will choose the corresponding vertex in A. Thus the chosen pair of vertices will be the same color and either both adjacent, or both not adjacent to the other chosen pair. Thus Player II can always preserve the partial isomorphism.   ∎

4.8   Vertex Refinement Corresponds to $\mathcal{C}_2$

It turns out that the expressive power of $\mathcal{C}_2$ is characterized by the well known method of vertex refinement (see [2,16]). Let $G = \langle V, E, C_1, \ldots, C_r \rangle$ be a colored graph in which every vertex satisfies exactly one color relation. Let $f : V \rightarrow \{1 \ldots n\}$ be given by $f(v) = i$ iff $v \in C_i$. We then define f', the refinement of f as follows: The new color of each vertex, v, is defined to be the following tuple:

$$\langle f(v), n_1, \ldots, n_r \rangle$$

where n_i is the number of vertices of color i that v is adjacent to. We sort these new colors lexicographically and assign $f'(v)$ to be the number of the new color class which v inhabits. Thus two vertices are in the same new color class just if they were in the same old color class, and they were adjacent to the same number of vertices of each color. We keep refining the coloring until at some level $f^{(k)} = f^{(k+1)}$. We let $\bar{f} = f^{(k)}$ and call $\bar{f}$ the *stable refinement* of f.

The equivalence of stable colorings and $\mathcal{C}_2$ equivalence is summed up by the following

Theorem 4.8.1 *Given a colored graph,* $G = \langle V, E, C_1, \ldots, C_r \rangle$, *with two vertices,* g_1 *and* g_2, *the following are equivalent:*

- *1.* $\bar{f}(g_1) = \bar{f}(g_2)$

- *2. For all* $\varphi(x_1) \in \mathcal{C}_2$, $G \models \varphi(g_1)$ *iff* $G \models \varphi(g_2)$.

- *3. Player II wins the* $\mathcal{C}_2$ *game on two copies of* G, *with pebble pair number 1 initially placed on* g_1 *and* g_2 *respectively.*

Proof By induction on r we show that the following are equivalent:

1. $f^{(r)}(g_1) = f^{(r)}(g_2)$

2. For all $\varphi(x_1) \in \mathcal{C}_2$ of quantifier depth r, $G \models \varphi(g_1)$ iff $G \models \varphi(g_2)$.

3. Player II wins the r move $\mathcal{C}_2$ game on two copies of G, with pebble pair number 1 initially placed on g_1 and g_2 respectively.

The base case is by definition. $f^{(0)}(g_1) = f(g_1) = f(g_2)$ iff g_1 and g_2 satisfy the same initial color predicate. This is true if and only if g_1 and g_2 satisfy all the same quantifier free formulas. This in turn is true if and only if the map sending g_1 to g_2 is a partial isomorphism. This last is the definition of Player II winning the 0 move game.

Assume that the equivalence holds for all g_1 and g_2 and for all $r < m$.

$(\neg 1 \Rightarrow \neg 2)$: Suppose that $f^{(m)}(g_1) \neq f^{(m)}(g_2)$. There are two cases. If $f^{(m-1)}(g_1) \neq f^{(m-1)}(g_2)$ then by the inductive assumption there is a quantifier depth $m - 1$ formula $\varphi \in \mathcal{C}_2$ on which g_1 and g_2 differ. Otherwise it must be that g_1 and g_2 have a different number of neighbors of some $f^{(m-1)}$ color class i. Let N be the maximum of these two numbers. By induction two vertices are in the same $f^{(m-1)}$ color class iff they agree on all quantifier depth $m-1$ $\mathcal{C}_2$ formulas. Since quantifier depth $m-1$ formulas are closed under conjunction and the graphs in question are finite there is a depth $m - 1$ $\psi_i \in \mathcal{C}_2$ such that for all $g \in G$,

$$f^{(m-1)}(g) = i \quad \Leftrightarrow \quad G \models (\psi_i)_g^{x_1}$$

It follows that g_1 and g_2 differ on the formula:

$$(\exists N x_2)(E(x_1, x_2) \wedge \psi_{i x_2}^{x_1}) .$$

$(\neg 2 \Rightarrow \neg 3)$: Suppose that $G \models \varphi_{g_1}^{x_1}$ but $G \models \neg \varphi_{g_1}^{x_1}$, for some $\varphi \in \mathcal{C}_2$ of quantifier depth m. If φ is a conjunction then g_1 and g_2 must differ on at least one of the conjuncts, so we may assume that φ is of the form $(\exists Nx_2)\psi(x_2)$. On the first move of the game Player I chooses the N vertices v such that $\psi(v)_{g_1}^{x_1}$. Whatever Player II chooses as B there will be at least one vertex v_2 such that $G \models \neg\psi(v_2)_{g_1}^{x_1}$. Player I puts his pebble number 2 on this v_2. Player II must respond with some $v_1 \in A$. The vertices v_1, v_2 now differ on a quantifier depth $m - 1$ formula. Thus by induction Player II loses the remaining $m - 1$ move game.

$(1 \Rightarrow 3)$:   Suppose that $f^{(m)}(g_1) = f^{(m)}(g_2)$. It follows that g_1 and g_2 have the same number of neighbors of each $f^{(m-1)}$ color. Thus a 1:1 correspondence exists between the vertices in the first copy of G and those in the second preserving both the property of being adjacent to g_i and the $f^{(m-1)}$ color. (Note that since we are considering two copies of the same graph, if both copies have the same number of red neighbors of g_i then they also both have the same number of red non-neighbors of g_i.) It follows that Player II can assure that after the first move the pair of vertices chosen will be in the same $f^{(m-1)}$ color class. Thus by the induction hypothesis Player II has a win for the remaining $m - 1$ move game.   ∎

ALL TREES AND ALMOST ALL GRAPHS

Theorem 4.8.1 combined with some facts about stable colorings provide us with several corollaries concerning graphs characterized by C_2. First, it is well known that the set of finite trees is characterized by stable coloring [1]. Thus:

Corollary 4.8.2 *Let TREES be the set of finite trees. Then vc(TREES) = 2.*

It is interesting to compare Corollary 4.8.2 with the more complicated situation in which counting is not present:

Fact 4.8.3 *[25] Let T_k be the set of finite trees such that each node has at most k children, and let S_k be the subset of T_k in which each non-leaf has exactly k children. Then,*

  1.

$$var(T_k) = \begin{cases} 2 & \text{if } k = 1 \\ 3 & \text{if } 2 \leq k \leq 3 \\ k & \text{if } k > 3 \end{cases}$$

  2.

$$var(S_k) = \begin{cases} 2 & \text{if } 1 \leq k \leq 2 \\ 3 & \text{if } 3 \leq k \leq 6 \\ \lceil k/2 \rceil & \text{if } k > 6 \end{cases}$$

Babai and Kučera have proved the following result about stable colorings of random graphs:

Fact 4.8.4 *[4] There exists a constant $\alpha < 1$ such that if G is chosen randomly from the set of all labeled graphs on n vertices then*

$$Prob\{G \text{ has two vertices of the same stable color}\} < \alpha^n .$$

Corollary 4.8.5 *Almost all finite graphs are characterized by C_2.*

It is easy to see that Fact 4.8.4 fails for regular graphs: all regular graphs of degree d on n vertices are C_2 equivalent. More recently, Kučera has given a linear algorithm for canonization of regular graphs of a given, fixed degree [26]. It follows from his results that:

Corollary 4.8.6 *For all d, and sufficiently large n, C_3 characterizes more than $1 - O[1/n]$ of the regular graphs of degree d on n vertices.*

4.9   Equivalence and Canonization Algorithms

The stable coloring of a graph is computable in $O[|E| \log n]$ steps [16]. We present the algorithm for completeness.

Algorithm 4.9.1 *1. Place indices $1, \ldots, r$ of initial color classes on list L.*
   2. **While $L \neq \emptyset$ do begin**
   3.   **For** *each vertex v adjacent to some color classes in L,*
            record how many neighbors of each such color class v has.
   4.   *Sort these records to form new color classes.*
   5.   *Replace L with indices of all but the largest piece of each old class.*

Theorem 4.9.2 *Algorithm 4.9.1 computes the vertex refinement of a graph G. It can be implemented to run in $O[|E| \log n]$ time on a RAM.*

Proof If we implement line 4 as a bucket sort then the amount of work in performing an iteration of the while loop is proportional to the number of edges traversed. Note that each time an edge is traversed, the color class of its head is at most half of its previous size. Thus $O[|E| \log n]$ steps suffice. ∎

Corollary 4.9.3 *We can test if $G \equiv_{C_2} H$ in $O[|E| \log n]$ steps, where $|E|$ is the number of edges in G.*

Proof We compute the stable coloring of $G \cup H$. G and H are C_2 equivalent iff each color class has the same number of vertices from each graph. ∎

As promised, we show how to modify the above algorithm to compute canonical labelings of graphs characterized by C_2.

Theorem 4.9.4 *Let S be a set of finite graphs characterized by $\mathcal{C}_2$. Then canonical labelings for S are computable in $O[|E| \log n]$ steps.*

Proof   We modify Algorithm 4.9.1 as follows: When a stable coloring is reached, if each vertex has a unique stable color, then a canonical labeling is determined. Otherwise, let C_i be the first color class of size greater than one, and let g be a vertex of color C_i. Make g a new color, C_{new}, add C_{new} to L and continue the refinement.

Suppose that $G, g \equiv_{\mathcal{C}_2} H, h$. Let G' and H' be the result of coloring g and h 'new'. Since $\mathcal{C}_2$ characterizes S, G' and H' are isomorphic. It follows that $\mathcal{C}_2$ equivalent graphs will result in the same canonical labeling. Furthermore, the analysis of the revised algorithm is unchanged.   ∎

We will next present the algorithm to test $\mathcal{C}_{k+1}$ equivalence for $k \geq 2$. Define *stable colorings of k tuples* as follows: Initially we give each k tuple of vertices from G a color according to its isomorphism type. That is $\langle g_1 \ldots g_k \rangle$ has the same initial color as $\langle h_1 \ldots h_k \rangle$, just if the map $\alpha : g_i \mapsto h_i, \quad i = 1 \ldots k$ is an isomorphism.

We next form the new color of $\langle g_1 \ldots g_k \rangle$ as the tuple:

$$\left\langle f(g_1 \ldots g_k), \mathrm{SORT}\Big\{ \langle f(g, g_2 \ldots, g_k), \ldots, f(g_1, \ldots, g_{k-1}, g) \rangle \mid g \in G \Big\} \right\rangle$$

That is the new color of a k-tuple is formed from the old color, as well as from considering, for each vertex g, the old color of the k k-tuples resulting from the substitution of g into each possible place.

Theorem 4.9.5 *A stable coloring of k tuples in an n vertex graph may be computed in $O[k^2 n^{k+1} \log n]$ steps.*

Proof   This is a generalization of Algorithm 4.9.1. We must refine the coloring for each color class, B_i, of k-tuples. Each such refinement takes $O[kn]$ steps for each k-tuple in B_i. Each of the n^k k-tuples will have its color class treated at most $\log(n^k)$ times.   ∎

Theorem 4.9.6 *Let G be a graph whose $k-1$ tuples of vertices are colored. Let $\vec{g}, \vec{h} \in G^{k-1}$. The following are equivalent.*

1. *$\bar{f}(\vec{g}) = \bar{f}(\vec{h})$*

2. *For all $\varphi(x_1 \ldots x_{k-1}) \in \mathcal{C}_k$, $G \models \varphi(\vec{g})$ iff $G \models \varphi(\vec{h})$*

3. *Player II wins the $\mathcal{C}_k$ game on two copies of G with pebbles $1 \ldots k-1$ initially placed on $g_1 \ldots g_{k-1}$ and $h_1 \ldots h_{k-1}$ respectively.*

Proof   The proof is similar to that of Theorem 4.8.1.   ∎

Corollary 4.9.7 C_k *equivalence may be tested in* $O[n^k \log n]$ *steps. (If* k *is allowed to vary with* n *this becomes* $O[k^2 n^k \log n]$.) *Similarly, if* S *is characterized by* C_k, *then canonical labelings for* S *may be computed in the same time bound.*

4.10   Conclusions

We have begun a study of which sets of graphs are characterized by the languages $\mathcal{L}_k$ and C_k. For such sets of graphs we have given simple and efficient canonization algorithms. General directions for further study include the following:

1. There are many interesting questions concerning the values $var(S)$ and $vc(S)$ for various classes of graphs S. In particular it would be very interesting to determine vc(Planar Graphs) and vc(Genus k) graphs.

2. Question 4.3.3 in its new form, "What must we add to first-order logic with fixed point and counting in order to obtain all polynomial-time graph properties" deserves considerable further study, cf. [8].

3. Fact 4.3.4 implies that (FO + LFP + counting) does not even include all of DSPACE[$\log n$]. It would be very interesting, and perhaps more tractable to answer question 2 for other classes such as NSPACE[$\log n$].

Acknowledgements: Thanks to Steven Lindell for suggesting some improvements to this paper.

4.11   REFERENCES

[1] A.V. Aho, J.E. Hopcroft and J.D. Ullman (1974), *The Design and Analysis of Computer Algorithms*, Addison- Wesley.

[2] Laszlo Babai, "Moderately Exponential Bound for Graph Isomorphism," Proc. Conf. on Fundamentals of Computation Theory, Szeged, August 1981.

[3] L. Babai, W.M. Kantor, E.M. Luks, "Computational Complexity and the Classification of Finite Simple Groups," *24th IEEE FOCS Symp.*, (1983), 162-171.

[4] Laszlo Babai and Luděk Kučera (1980), Canonical Labelling of Graphs in Linear Average Time," *20th IEEE Symp. on Foundations of Computer Science*, 39-46.

[5] Laszlo Babai and Eugene M. Luks, "Canonical Labeling of Graphs," *15th ACM STOC Symp.*, (1983), 171-183.

[6] D. Mix Barrington, N. Immerman, and H. Straubing, "On Uniformity Within NC^1," *Third Annual Structure in Complexity Theory Symp.* (1988), 47-59.

[7] Jon Barwise, "On Moschovakis Closure Ordinals," J. Symb. Logic 42 (1977), 292-296.

[8] J. Cai, M. Fürer, N. Immerman, "An Optimal Lower Bound on the Number of Variables for Graph Identification," *30th IEEE FOCS Symp.* (1989), 612-617.

[9] Ashok Chandra and David Harel, "Structure and Complexity of Relational Queries," *JCSS* **25** (1982), 99-128.

[10] A. Ehrenfeucht, "An Application of Games to the Completeness Problem for Formalized Theories," *Fund. Math.* **49** (1961), 129-141.

[11] Ron Fagin, "Probabilities on Finite Models," *J. Symbolic Logic* **41**, No. 1 (1976), 50-58.

[12] R. Fraissé, "Sur les Classifications des Systems de Relations," *Publ. Sci. Univ. Alger* **I** (1954).

[13] Leslie Goldschlager, "The Monotone and Planar Circuit Value Problems are Log Space Complete for P," *SIGACT News* **9**, No. 2 (1977).

[14] Yuri Gurevich, "Logic and the Challenge of Computer Science," in *Current Trends in Theoretical Computer Science*, ed. Egon Börger, Computer Science Press.

[15] Christoph M. Hoffmann, *Group-Theoretic Algorithms and Graph Isomorphism*, Springer-Verlag Lecture Notes in Computer Science 136 (1982).

[16] John E. Hopcroft and Robert Tarjan, "Isomorphism of Planar Graphs," in *Complexity of Computer Computations*, R. Miller and J.W Thatcher, eds., (1972), Plenum Press, 131-152.

[17] Neil Immerman, "Number of Quantifiers is Better than Number of Tape Cells," *JCSS* **22**, No. 3, June 1981, 65-72.

[18] Neil Immerman, "Upper and Lower Bounds for First Order Expressibility," *JCSS* **25**, No. 1 (1982), 76-98.

[19] Neil Immerman, "Relational Queries Computable in Polynomial Time," *Information and Control,* 68 (1986), 86-104.

[20] Neil Immerman, "Languages That Capture Complexity Classes," *SIAM J. Comput.* **16**, No. 4 (1987), 760-778.

[21] Neil Immerman, "Nondeterministic Space is Closed Under Complementation," *SIAM J. Comput.* **17**, No. 5 (1988), 935-938.

[22] Neil Immerman, "Expressibility and Parallel Complexity," *SIAM J. of Comput* **18** (1989), 625-638.

[23] Neil Immerman, "Expressibility as a Complexity Measure: Results and Directions," *Second Structure in Complexity Theory Conf.* (1987), 194-202.

[24] Neil Immerman, "Descriptive and Computational Complexity," in *Computational Complexity Theory,* ed. J. Hartmanis, *Proc. Symp. in Applied Math.,* 38, American Mathematical Society (1989), 75-91.

[25] Neil Immerman and Dexter Kozen, "Definability with Bounded Number of Bound Variables," *Information and Computation,* **83** (1989), 121-139.

[26] Luděk Kučera, "Canonical Labeling of Regular Graphs in Linear Average Time," *28th IEEE FOCS Symp.* (1987), 271-279.

[27] Eugene M. Luks, "Isomorphism of Graphs of Bounded Valence Can be Tested in Polynomial Time," JCSS 25 (1982), pp. 42-65.

[28] Yiannis N. Moschovakis, *Elementary Induction on Abstract Structures,* North Holland, 1974.

[29] Bruno Poizat, "Deux ou trois chose que je sais de Ln," *J. Symbolic Logic,* 47 (1982), 641-658.

[30] Simon Thomas, "Theories With Finitely Many Models," *J. Symbolic Logic,* **51**, No. 2 (1986), 374-376.

[31] M. Vardi, "Complexity of Relational Query Languages," *14th Symposium on Theory of Computation,* 1982, (137-146).

5

Self-Reducibility: Effects of Internal Structure on Computational Complexity

Deborah Joseph[1]
Paul Young[2]

ABSTRACT In this paper we discuss the effect that *self-reducibility* properties have on the analysis of complexity classes. We begin by reviewing some of the more elementary results for readers unfamiliar with the field, and then we discuss some recent results and directions where self-reducibilities have been useful. Throughout, we focus on the question of when self-reducibility properties cause sets, or classes of sets, to have lower complexity than might otherwise be expected. This paper is an attempt to provide an overview of known results and suggest unifying concepts. By doing so we suggest that *a continuing systematic study of the relationship between the internal structure of a set and the computational complexity of a set is in order.*

5.1 Introduction

Structural complexity theory is often concerned with the inter-relationship between sets in a complexity class, (e.g. *Is set A complete for class C?* or *Are all* NP-*complete sets polynomially isomorphic?*), and inclusion relationships between complexity classes, (e.g. *Does* P = NP*?*). Structural complexity theorists have also studied the internal properties of sets, (e.g. *Do all* NP-*complete sets have polynomially decidable subsets?*). It is this later type of internal structure that we address in this paper. In particular, we will be discussing the effect that *self-reducibility* properties have on

[1]Computer Sciences Department, University of Wisconsin, Madison, WI.
[2]Computer Science Department, University of Washington, Seattle, WA.

This paper combines material presented at the Second Annual Structures in Complexity Theory Conference in a paper coauthored with Judy Goldsmith with material from a guest column that we wrote for Juris Hartmanis' Structural Complexity column in the Bulletin of the EATCS. This work was supported in part by NSF grant DCR-8520597 and by a Brittingham Visiting Professorship from the University of Wisconsin.

computational complexity. This line of research is not new. In fact, viewed broadly, the notion of self-reducibility has been one of the more pervasive notions in complexity theory. It dates back to early work by Trachtenbrot, Schnorr, Selman, and Meyer and Paterson, and it has continued to be an important tool. For instance it recently played a key role in Krentel's results on functional hierarchies for optimization problems and in Beigel and Wagner's work on query hierarchies. Self-reducibility also plays a critical role in Kadin's proof that if the Boolean hierarchy over NP collapses then the polynomial time hierarchy collapses.

A set is *self-reducible* if the membership question for any element can be reduced in polynomial time to the membership question for a number of *shorter* elements. The classic example of a self-reducible set is SAT, the set of satisfiable Boolean formulas. SAT is self-reducible because an arbitrary Boolean formula $B(x_1, x_2, \ldots, x_n)$ is satisfiable *if and only if* at least one of the two *shorter* Boolean formulas $B(0, x_2, \ldots, x_n)$ or $B(1, x_2, \ldots, x_n)$ is satisfiable.

The potential importance of the self-reducibility of SAT and of other sets was recognized early, and in the 1970's the property was defined, generalized and investigated by Trachtenbrot, Schnorr, Selman, and Meyer and Paterson, ([Tra70,Sch76,Sch79,Sel79,MP79]). Nice reviews of this early work can be found in Balcázar ([Bal90a]), Wagner and Wechsung ([WW86]), Selman ([Sel88a]), and Mahaney ([Mah86,Mah89]), the latter particularly for connections with sparse sets. In this paper we will be primarily interested in reviewing some of the more elementary results for readers unfamiliar with the field and then discussing some recent results and directions where self-reducibilities have been useful. Throughout, we will focus on the question of what effect self-reducibilities have on the computational complexity of a set.

We first discuss work going back to Berman, Fortune, Mahaney, Karp-Lipton, Long, and Yap. This work uses the fact that certain sets are both self-reducible and complete for various levels of the polynomial time-hierarchy to show how reductions to sparse sets can force collapses in the polynomial time hierarchy. Of the exponentially many elements of size n, sparse sets, by definition, contain only polynomially many elements. Thus, results of this form are a strong indication that the information in complete sets cannot be compressed into a polynomial amount of information.

Next we discuss a generalization of the notion of Turing self-reducibility: *self-helping*. This is followed by a discussion of more recent work by Selman and Balcázar which, following up early work by Schnorr, uses self-reducibilities to show that all functional versions of NP-complete problems are Turing reducible in polynomial time to their corresponding decision problems. Moving further in the direction of classifying functional problems, the self-reducibility of SAT has been used by Krentel, by Wagner, and by Beigel to show that certain functions can be computed in polynomial time by a machine that makes a small number of oracle queries to SAT.

Finally, we look in still another direction. Beginning with work by Selman showing that a set is in P *if and only if* it is both *disjunctive self-reducible* and *p-selective,* recent work has shown that self-reducibility properties which are generally expected to define classes broader than P work in combination to give precise, and sometimes unexpected, characterizations of P. Here we discuss more general notions of self-reducibility including *near-testability* and *p-cheatability,* and we discuss how these properties can combine to restrict a set's complexity.

5.2   Elementary Applications

A set A is *Turing self-reducible* (or just *self-reducible* for short) if there is a polynomial time oracle machine M^A that uses A itself as an oracle to recognize A, with the restriction that on inputs of length n all of M^A's queries are required to have *length* less than n.[3] Obviously, all *polynomially* decidable sets are self-reducible in this sense, and all sets that are self-reducible are decidable in exponential time (time 2^{poly}).

If we consider the case of SAT, then the self-reducibility mentioned earlier,

$$B(x_1, x_2, \ldots, x_n) \in SAT \iff$$

$$[\ B(0, x_2, \ldots, x_n) \in SAT \ \text{ or } \ B(1, x_2, \ldots, x_n) \in SAT \ ],$$

yields, in an obvious way, a disjunctive *self-reduction tree.* To see this begin by placing $B(x_1, x_2, \ldots, x_n)$ at the root of the tree and then give each node two daughter nodes by fixing the first free variable first at 0 ("false"), and then at 1 ("true"). The depth of this tree is n, which corresponds to the length of the possible satisfying assignments, while the number of leaves in the tree is 2^n, which corresponds to the number of possible satisfying assignments. The Boolean formula $B(x_1, x_2, \ldots, x_n)$ at the root of the tree is satisfiable just if, for each level of the tree, at least one node at that level is satisfiable. Furthermore, since the leaves of the tree have no free variables, but simply have assignments of "true" or "false" to the variables, each leaf of the tree can be tested for satisfiability in linear time.

An exponential time (and polynomial space) algorithm for SAT can thus be obtained from the tree by doing a depth-first search of the tree and

---

[3]Meyer and Paterson were the first to define self-reducibility as we use it today. To obtain full generality and to preserve the concept under polynomially computable isomorphisms, they use any polynomially computable, polynomially well-founded partial order to measure the "length" of strings. For the purposes of exposition, using ordinary lengths of binary strings is adequate.

using the fact that the formula, $B(x_1, x_2, \ldots, x_n)$, at the root of the tree is satisfiable just if at least one of the leaves of the tree is satisfied. Furthermore, one can terminate the depth-first search at the first leaf which belongs to SAT.

This by itself is not very interesting, but if SAT is reducible to *any* set by a polynomial time computable function g, then, because the self-reducibility is *disjunctive,* the collisions among the values $g(B_i)$ can be used to *prune* the depth-first search tree: if two nodes of the tree B_i and B_j both map to the same element, i.e., if $g(B_i) = g(B_j)$, then B_i is satisfiable *if and only if* B_j is satisfiable. Thus if B_j lies on a branch to the right of the branch containing B_i *there is no reason to continue the depth-first search if the search reaches node B_j*: if B_j is satisfiable the depth-first search would already have discovered that node B_i is satisfiable. Thus we can safely *prune* the subtree beginning at node B_j from the depth-first search tree when we reach node B_j. The net result is a depth-first search tree for a formula B of length n that has the property that at *any* level of the tree the number of *unsatisfiable* nodes at that level cannot exceed the cardinality of the set $g(\overline{SAT} \cap \{x : |x| \leq n\})$.[4] Thus, if the entire range of g could ever be forced to be *polynomially sparse,* that is, if

$$|\{g(x) : |x| \leq n\}| \leq b(n)$$

for some polynomial b, then the pruned depth-first search tree would have at most $b(n)$ leaves, and so there would be a polynomial time algorithm for SAT.

The above ideas formed the basis for the early work by Piotr Berman, Fortune, and Meyer and Paterson, ([Ber78] [5], [For79] and [MP79]), concerning self-reducibility and reductions to sparse sets. In fact, as they observed, if we merely had that $g(\overline{SAT})$ is sparse, the pruned tree would have at most $b(n)$ *unsatisfied leaves,* and so the depth-first search algorithm would terminate after visiting at most $b(n) + 1$ leaves of the pruned tree. Thus, in the case where SAT is reducible by g to a *co-sparse* set, the depth-first search tree can safely be pruned to a polynomial sized tree by pruning nodes which map under g to a common value and by terminally pruning the depth-first search tree after it has reached only $b(n) + 1$ leaves. Therefore, if $g(\overline{SAT})$ is sparse there is a polynomial time decision procedure for SAT.

---

[4]We will frequently be considering the finite initial subsets of a set; for instance, $\overline{SAT} \cap \{x : |x| \leq n\}$. For notational simplicity we henceforth denote such a set by $\overline{SAT}_{\leq n}$.

[5]P. Berman's work is actually based on the more complicated problem of $CLIQUE$, but it developed many of the ideas that we have described above for SAT.

For later use, note first that *any* pruning of the self-reducibility tree for
SAT to a polynomial sized tree yields a polynomial time decision procedure
for *some subset* of *SAT,* provided our acceptance rule is that we accept the
formula at the root of the tree just if at least one of the leaves is verified to
be a satisfying instance of the formula. Note also that to work correctly on
formulas of length n, g only needs to be a correct reduction for all formulas
of length $< n$. [6]

With Berman and Fortune's result, we see that, unless P = NP, no co-
sparse set can be *hard* for NP under polynomially computable *functional*
reductions. Now if we instead had that $g(SAT)$ is polynomially sparse,
we would know that the pruned depth-first search tree had at most $b(n)$
satisfied leaves. Thus, after visiting at most $b(n) + 1$ leaves of the pruned
tree, the depth-first search algorithm would have visited an *unsatisfied* leaf.

---

[6] Osamu Watanabe made the nice observation that breadth first search may
be used as a tree traversal strategy in Fortune's proof. His argument can be
summarized as follows. Let $b(n)$ be a polynomial bounding function as above. As
before, to decide whether a Boolean formula B, of length n, is in *SAT* we look
at its self-reduction tree. However, this time we do a breadth first search of the
tree.

We search the tree level by level. As we search a level we build up a list of
nodes whose children will be searched at the next level. For the algorithm to run
in polynomial time we need only make sure that at each level this list contains
only a polynomial number of nodes. For this purpose we keep a counter, which
is set to 0 at the beginning of the search over each level.

Searching a level: When a node x is visited we compute $g(x)$. If $g(x)$ is distinct
from any other element in the range of g that we have found *at this level*, then we
increment the counter and add $x's$ children to the list of nodes to be searched at
the next level. (If $g(x)$ is not distinct, then we do nothing and thus x's children
will not be visited when we seach the next level. This is the pruning that is done.)
If the counter value is greater than $b(n)$, then we stop the entire search, otherwise
we consider the next node on the level. When a level is completed we go on to
the next level, resetting the counter.

The search can terminate in one of two ways: 1) it reaches the bottom level of
the tree, or 2) the counter causes it to stop earlier.

Case 1: the pruning strategy used in this algorithm is the same as that described
above and thus if the bottom level of the tree is reached, then it is easy to verify
whether or not B is in *SAT*.

Case 2 (the interesting case): if the counter is exceeded, then a large number
of elements in the range of g were found. In particular, more elements were found
than could possibly be in the sparse set, S. Therefore at least one of the nodes
at the level where the search terminated mapped to an element in $\overline{S}$. That node
is therefore in *SAT* . (Remember, *SAT* maps to $\overline{S}$ and $\overline{SAT}$ maps to S.) But, if
that node is in *SAT*, then so is B. Thus, whenever the search terminates because
the counter is exceeded the formula B is in *SAT*.

Notice two more things. First, if the search terminates because the counter is
exceeded then, the argument given above does not provide a satisfying assignment
(witness) for the formula. Second, if there are no inputs for which the search
terminates because of the counter, then *SAT* is also reducible to a sparse set.

But, since a test for *satisfiability* cannot halt when it simply reaches an *un*satisfied leaf, this case does *not* yield a polynomial time algorithm for *SAT*.

Nevertheless, this situation deserves closer scrutiny. For this case, Mahaney, ([Mah82]), gave a method for "estimating" (or guessing) an "optimal" pruning of the self-reducibility tree. This is a clever technique, which can be expected to have wide applicability. For example, it is one historical root for the technique used by Immerman and by Szelepcsényi in their recent independent work proving the closure of various nondeterministic space classes under complementation, ([Imm88,Sze88]).

When Mahaney proved that if *SAT* is $\leq_m^P$-reducible to a sparse set S, then P = NP, his *proof* required that S be in NP. In the same paper he pointed out how, given a reduction of a set A in NP to an arbitrary sparse set S, to give a reduction of A to a sparse set $S\#$ *in* NP. However, this seems often to be overlooked, and in the subsequent literature, Mahaney's theorem is frequently misquoted as requiring that the sparse set S be *in* NP. It is thus instructive to modify Mahaney's (and Fortune's) proof(s) to directly obtain the full version of Mahaney's theorem.

Theorem 5.2.1 ([Mah82]) P = NP *if and only if there exists a sparse set S such that SAT $\leq_m^P$ S.*

For the proof, we begin by considering Fortune's *proof* discussed above. This time however, we have that $SAT \leq_m^P S$ where S rather than $\overline{S}$ is sparse, and instead of bounding $|g(\overline{SAT}_{\leq n})|$, the bounding function b now insures that

$$|g(SAT_{\leq n})| \leq b(n).$$

Changing Mahaney's definition slightly, we define the *pseudo-complement* of *SAT* as

$$PC(SAT) =_{def} \{\langle x, m \rangle \ : \ m \leq b(|x|) \text{ and } (\exists_{distinct} \ g(y_1), \ldots, g(y_m))$$
$$(\forall i) \ [y_i \in SAT_{\leq |x|} \text{ and } g(y_i) \neq g(x)]\}.$$

Clearly, $PC(SAT) \in NP$.

Before we describe the pruning algorithm we should consider the relationship between $PC(SAT)$ and $\overline{SAT}$. To do so, for each n and for each $m \leq b(n)$ define

$$PC(SAT)_{m,n} =_{def} \{x : |x| \leq n \text{ and } \langle x, m \rangle \in PC(SAT)\}.$$

Now observe that if $m_0 = |g(SAT_{\leq n})|$, i.e., if m_0 happens to be the actual number of distinct elements of S which elements of *SAT* of length less than or equal to n map to under g, then $PC(SAT)_{m_0,n} = \overline{SAT}_{\leq n}$. I.e., since

$PC(\overline{SAT})$ is in NP, the "optimal guess" m_0, of the value m "seems" to place $\overline{SAT}$ into NP. [7]

Now since SAT is complete for NP,

$$PC(SAT) \ \leq^{\mathrm{P}}_m \ SAT \ \leq^{\mathrm{P}}_m \ S,$$

so let f be a polynomial time computable function which reduces $PC(SAT)$ to S.

Let b' be a polynomial bounding function, derivable from the sparsity of S, from b, and from f satisfying

$$\mid \{f(x,m) : |x| \leq n \text{ and } m \leq b(|x|)\} \ \cap \ S \mid \ \leq \ b'(n).$$

Next, for any fixed n, for all x of length $\leq n$, consider each of the reductions

$$\lambda x f(x,m) \quad \text{for constants } m \text{ with } m \leq b(|x|).$$

For each of these $b(|x|) + 1$ reductions, we can prune the self-reducibility tree for SAT by using the reduction $\lambda x f(x,m)$ to prune the branches of the tree and stopping the generation of the tree once it has $b'(|x|) + 1$ leaves. This yields a polynomial number of pruned self-reduction trees, each of polynomial size. So we can test them all in polynomial time. We have already seen that for each n the "optimal" guess m_0, which guesses $|g(\overline{SAT}_{\leq n})|$, actually yields (on all formulas of length $\leq n$) a reduction of $\overline{SAT}$ to the sparse set S. Thus, from the proof of Fortune's theorem, we see that for *this* choice of m the pruned reduction tree must actually decide SAT in polynomial time. While the other pruned reduction trees cannot be expected to decide SAT, we have already remarked that a pruned reduction tree for SAT can never yield an answer "yes" unless the root of the tree really is in SAT. Thus a root element is in SAT *if and only if* at least one of these polynomially many pruned reduction trees claims that it *is* in SAT.[8]

This shows that if SAT is $\leq^{\mathrm{P}}_m$ reducible to a sparse set, then P = NP. The proof of the converse is trivial, since in this case SAT would be reducible in polynomial time to *any* set except the empty set and the universal set.

A number of authors, including Ukkonen ([Ukk83]), Yap ([Yap83]), and Yesha ([Yes83]), have extended the Fortune – Mahaney collapse of NP to P. For example, Yesha shows that if there is a sparse set S which is *hard* for

---

[7]Intuitively, the essential difference between our proof and Mahaney's is that, while Mahaney guessed the census of S, we guess the census of $g(SAT)$. Since the latter is in NP, while the former may not be, his proof needed that $S \in$ NP.

[8]It is worth noting that even incorrect guesses of m_0 may result in pruned trees which provide a correct "yes" answers about B's membership in SAT. That is, if B is in SAT , such a tree may contain a path which corresponds to a satisfying assignment.

co-NP or *complete* for NP under a *positive* polynomial time bounded-truth-table reducibility, then P = NP. Given the current interest in the Boolean hierarchy (the closure of NP under polynomial time bounded-truth-table reductions), it would be particularly interesting to know whether such results can be extended to *all* bounded-truth-table reductions.[9] We should note here that Book and Ko, ([BK87]), have shown that the collection of all sets $\leq_{k-tt}^{P}$ reducible to a sparse set is properly contained in the collection of all sets $\leq_{k+1-tt}^{P}$ reducible to a sparse set, and that these separations can be witnessed by sets which are in *exponential time*. An extension of the results of Fortune – Mahaney – Yesha to all *bounded-truth-table* reductions would show that the Book – Ko separations can not be accomplished by sets which are in the Boolean hierarchy. (Further separation results for classes reducible to sparse sets may be found in [Ko88].)

Obviously, the Fortune – Mahaney results apply to sets which are *hard* for NP or for co-NP under $\leq_{m}^{P}$-reductions. But another interesting open question, pointed out by Mahaney in [Mah86], is the extent to which the Fortune – Mahaney results apply to sets *in* NP − P which are *not* hard for NP. The Fortune proof will clearly apply to *any* set which is disjunctive self-reducible, but Mahaney's proof seems to require *both* disjunctive self-reducibility *and* completeness of the set reducible to the sparse set; i.e., the sparse set must be *hard* for NP. As Mahaney points out, *GRAPH-ISOMORPHISM* remains a particularly interesting problem. This problem is easily seen to be disjunctive self-reducible. As shown in [MP79] (and as follows directly from Fortune's proof), the *complement* of *GRAPH-ISOMORPHISM* cannot be reduced to a sparse set without collapsing *GRAPH-ISOMORPHISM* to P . But because it is not known whether *GRAPH-ISOMORPHISM* is complete, in spite of its self-reducibility it still remains unknown whether *GRAPH-ISOMORPHISM* can be reduced to a sparse set.[10]

It is easy to see that every disjunctive self-reducible set is in NP, ([Ko83], so it is natural to ask whether *every* set in NP is disjunctive self-reducible. However, this is probably not true. In [Sel88a] Selman shows that for linear exponential time, if E $\neq$ NE, then NP − P contains a p-selective set. But, as we shall see later, any disjunctive self-reducible set which is $\leq_{m}^{P}$-equivalent to a p-selective set is in P .

---

[9]Very recently, working with Tim Long, we have shown that if *SAT* is *bounded-truth-table reducible* to a sparse set in co-NP, then the polynomial hierarchy collapses to NP ∩ co-NP, ([JLY89]).

[10]Recent work of Boppana, Håstad and Zachos on interactive proof systems shows that if *GRAPH-ISOMORPHISM* is NP-complete, then the polynomial time hierarchy collapses to $\Sigma_2^P \cap \Pi_2^P$, ([BHZ87]). Further work on *GRAPH-ISOMORPHISM*, making use of both self-reducibilities and *promise-problems*, can be found in [Sel88b].

The above work of Mahaney *et.al.* does not answer the question of whether sparse and co-sparse sets can be complete for NP under polynomial time Turing reductions. Various people have worked on this problem, ([KL80,Lon82,Mah82,Yap83,BBS86,Kad87]). Perhaps the strongest of these results is by Long, who proves the very elegant (and usually not fully quoted) result that if $SAT \leq_T^P S$ for a sparse set S, *then every set in the polynomial time hierarchy is* $\leq_T^P$-*reducible to this same sparse set, S.* It follows for example that if SAT is $\leq_T^P$ reducible to a sparse or co-sparse set in Δ_2^P, (i.e., if $\Delta_2^P = \{L : L \leq_T^P SAT\}$ has a sparse complete set), then the whole polynomial time hierarchy collapses to Δ_2^P. There is a good discussion of other similar results in [Mah89]. Mahaney also gives a nice discussion of how the disjunctive self-reducibility of SAT can be used to show that if SAT has polynomial sized circuits (i.e., is polynomial time Turing reducible to a sparse set), then there is a Π_1^P description of such circuits, and an excellent discussion of applications of this technique to many results on collapsing the polynomial time hierarchy. Mahaney also points out that not all of the original Karp – Lipton methods, ([KL80]), use the technique of self-reducibility, and he raises the question of whether self-reducibility can, in effect, be made a universal method.

Balcázar, Book, and Schöning have worked on this problem, and the results in [BBS86,Bal90a,Bal90b,BS90] show that all known results proving that reducibilities of complete sets to sparse sets force collapses in the polynomial time hierarchy can be obtained as a fairly *uniform* application of the self-reducibility technique.[11]

Finally, before we leave this section we shall briefly mention some of the recent work of Schöning, Balcázar, Ko, Hartmanis and Hemachandra on *robustness* and *helping* – notions which are closely related to Turing and truth-table self-reducibility.

These notions were introduced by Schöning and studied by Balcázar, Ko, Hartmanis and Hemachandra in [Sch85,Bal90b,Ko87] and [HH87]. Schöning defined an oracle Turing machine M to be *robust* if for every oracle A the language accepted by M^A is the same. Thus, changing the oracle can influence the resources used by the machine on an input (in fact, can change the computation), but it cannot change the acceptance or

---

[11]For example, in [Bal90a], Balcázar introduces the notion of *word decreasing query* (wdq) self-reducibility. A set A is wdq self-reducible if there is a polynomial time oracle machine M^A that uses A itself as oracle to recognize A, with the restriction that on an input x all of M's oracle queries must either be to strings shorter than x or to lexicographic predecessors of x. Balcázar observes that all wdq self-reducible sets are decidable in *exponential time* and shows that if a wdq self-reducible set is in PSPACE/*poly*, then it is in PSPACE. He then uses wdq self-reducible sets to reprove some of the Karp-Lipton results on the collapse of *exponential time* that were originally proved by the "round-robin tournament" method.

rejection of the input. With this definition Schöning showed the following.

Theorem 5.2.2 ([Sch85])
1. $L \in$ NP$\cap$ co-NP if and only if there is a robust, deterministic oracle Turing machine M and an oracle A such that M^A decides membership in L and M^A runs in polynomial time.
2. $L \in$ NP if and only if there is a robust, nondeterministic oracle Turing machine M and an oracle A such that M^A decides membership in L and M^A runs in polynomial time.

Now, returning to our claim that the notion of robustness is related to self-reducibility, consider the definition given by Ko of a *self-1-helper*.

Definition 5.2.3 ([Ko87]) *A is a self-1-helper if there is a robust, deterministic oracle Turing machine M accepting A such that for all $x \in A$, $M^A(x)$ halts in polynomial time.*

Ko shows that any disjunctive self-reducible set, for instance SAT, is a self-1-helper.

Theorem 5.2.4 ([Ko87]) *If A is disjunctive self-reducible, then A is a self-1-helper.*

The proof of this theorem rests on the fact that a robust, deterministic oracle Turing machine can be constructed which uses the oracle to aid in searching the self-reducibility trees for potential elements of A. The oracle quickly guides the algorithm to a leaf in the self-reduction tree, if the leaf element is in A, then the element at the root of the tree is also in A. If the leaf element is not in A, the algorithm can perform an exhaustive search of the self-reduction tree assure itself that the root element is not in A.

Recall that we mentioned above that it is not known whether all NP-complete sets are disjunctive self-reducible. Ko has observed that if every polynomial Turing degree in NP contains a disjunctive self-reducible set, then there are no log^*-sparse sets in NP $-$ P, and hence E $\not\subseteq$ NP. [12]
One of the major open problems left by Ko's work is: What is the relationship between self-helping and self-reducibility? For instance, is the converse of the above theorem true? If A is a self-1-helper, then it appears that A is close to being Turing self-reducible: the machine M^A decides membership in A in polynomial time, but what length queries are made to A? Unfortunately, there is no obvious reason that the fact that A is a self-1-helper should guarantee that there is an ordering of Σ^* relative to which all queries are length decreasing. In [Ko87] Ko discusses further conditions that can be added to the definition of self-1-helper so that this is true.

---

[12] The *polynomial time Turing degrees* in NP are the equivalence classes for the equivalence relation $\equiv_T^P$ in NP. ($A \equiv_T^P B$ *if and only if* $A \leq_T^P B$ *and* $B \leq_T^P B$.)

Balcázar, [Bal90b], made a further attempt to characterize the class of self-1-helpers. He said that a set $A \in$ NP has *self-computable witnesses* if two conditions hold. First, there is a set $B \in$ P such that

$$x \in A \iff (\exists y) \left[|y| \leq p(|x|) \text{ and } \langle x, y \rangle \in B \right]$$

for some polynomial p; that is, there is a polynomially recognizable set of short witnesses for A. Second, there is a function f that is polynomially computable relative to A that finds short witnesses for A; that is, if $x \in A$, then $\langle x, f(x) \rangle \in B$. Notice that SAT has self-computable witnesses and this follows directly from the disjunctive self-reducibility property of SAT. Balcázar proves the stronger fact that all self-1-helpers have self-computable witnesses.

Theorem 5.2.5 ([Bal90b]) *A set A is a self-1-helper if and only if it has self-computable witnesses.*

5.3   Self-reducibility and Functional Computation

Many problems which we traditionally think of as set recognition problems are more naturally thought of as functional computations, or clearly involve functional computations as the most obvious means of solution. For problems such as the Traveling Salesperson Problem, this is surely obvious from its statement, but even in other problems, for example clique problems, one is presumably more interested in knowing not just, e.g., whether there is a clique of a given size, but in knowing an actual instance of a maximal clique, or perhaps the maximum size of a clique. Even in asking whether the maximal clique of a graph is unique, which is a set-recognition problem, one expects that the difficulty of answering the question should have something to do with how hard it is to find a maximal clique.

In fact, Papadimitriou and Yannakakis, ([PY84]), show that the problem "Does the maximal clique have size *exactly k?*" is complete for the class D_2^P which is at the second level of the *Boolean* hierarchy over NP, and (presumably) well below Δ_2^P, which is the second level of the polynomial time hierarchy.

While showing that a problem is complete or *hard* for NP is generally believed to be adequate for showing that no fast (i.e., polynomial time) algorithms exist for the set, knowing the exact classification and structure of such sets is important because, as discussed by Papadimitriou and Yannakakis, ([PY88]), such results may have a bearing on whether efficient *approximation* or *probabilistic* algorithms will exist for the sets. As we shall see, Krentel (and others) have recently shown that understanding hierarchies of *functional* computations may yield useful insight into such *set* classification problems.[13]

While mathematically we are trained to think of functions as being reducible to sets, (namely each function is equivalent to its *graph)*, from the point of view of computations this notion of equivalence seems too gross. For example, given a method of computing a function, one can reduce the question of whether a pair $\langle x, y \rangle$ is in the graph of the function by making *a single* request for a computation of f. But the reverse question, given x asking what is the *value* of $f(x)$, is by no means easily related to asking membership questions about the graph of f. In the worst case, one might need $f(x)$ queries to the graph to compute $f(x)$. Even if one is clever and tries to code a function f, not by the graph of f, but instead by the more complicated set,

$$S_f =_{def} \{\langle x, y \rangle : y \text{ is an initial segment of } f(x)\}.$$

Then, even if one knows $|f(x)|$, it still seems to require $|f(x)|$ questions to S_f to compute $f(x)$.

It is interesting to observe that many of the functions one is interested in computing, for example, the minimal traveling salesperson tour or the maximal satisfying assignment, are self-reducible as *functions*. To date, this fact, the self-reducibility of *functional* computations, seems to have had little direct use in the literature. However, as we shall see, the self-reducibility of sets like *SAT* has been well-exploited in building *functional* hierarchies which help in analyzing set recognition problems.

Trachtenbrot, ([Tra70]), in his early work on *auto-reducibility*, was interested, not directly in the self-reducibility of *sets,* but rather in the question of when, in computing a function $f(x)$, knowing values of $f(y)$ for $y \neq x$ would help in the computation of $f(x)$. Because his definition did not require $y < x$, it led to questions concerning more general notions of self-reducibility such as *helping,* and *p-cheatability,* rather than the notion of Turing self-reducibility. However by the mid to late 1970's, Schnorr was interested in using self-reducibilities to analyze the difficulty of functional computations, ([Sch76,Sch79]). He showed that, for sets in NP which have a certain special form of disjunctive self-reducibility, there is a related functional problem for which the functional computation is "equivalent in difficulty" to the original decision problem.

---

[13] An overview of work in structural complexity theory on NP optimization problems can be found in [BJY90]. In that paper we discuss work which relates the internal structure of NP-complete sets to the polynomial time approximability of the corresponding NP optimization problems. (A small part of that work is also discussed in the following pages of this paper.) From work on NP optimization problems, it is clear that polynomial time isomorphisms do *not* generally preserve enough of the internal structure of sets to preserve polynomial time approximability. Restricted types of reductions which do seem to preserve enough such internal structure are discussed in the final section of [BJY90].

But as we have already seen, it is not likely that every set in NP is disjunctive self-reducible. Nevertheless, it is natural to associate with any set A in NP the set of all initial segments of the strings which "verify" that elements of A are in A, (for example, the initial segments of satisfying assignments for Boolean formulas). This set is obviously self-reducible, and one might expect it to be somehow equivalent to A. But Selman, ([Sel84]), gives a proof, credited to Borodin and Demers, ([BD76]), which shows that if P $\neq$ NP$\cap$ co-NP, then there is a set $A \in$ NP $-$ P for which the associated set of "verifying prefixes" is not even Turing reducible to A in polynomial time. In [Sel88a], Selman improves this to show that if P $\neq$ NP, then there is a set $A \in$ NP $-$ P for which the associated set of "verifying prefixes" is $\leq^{\mathrm{P}}_T$-complete for NP even though A itself is not complete for NP. On the other hand he proves that if A is $\leq^{\mathrm{P}}_T$-complete for NP, then A is polynomial Turing equivalent to its set of "verifying prefixes." In [Bal90b], generalizing Schnorr's results on the equivalence of certain NP problems and their functional counterparts to all NP-complete problems, Balcázar uses the self-reducibility of initial segments of verifying solutions in much the same way. For example, he shows that for an arbitrary NP-complete set A, a function, $f(x)$, verifying that $x \in A$ can always be computed in polynomial time relative to an oracle for A.

Balcázar's interpretation of this result is that for NP-complete problems there is always a functional solution which is "not much harder to compute than A itself." Obviously, except for the polynomial bound on the number of queries, this interpretation ignores the number of queries which the functional computation must make to the oracle A. Recently there has been much work developing hierarchies of functions which are based on the number of queries which the computations make to oracles. Some of this, for example the work of Beigel, ([Bei90]), shows that such hierarchies often differ significantly from the hierarchies of sets which are built by the same criteria.[14] Such work demonstrates the importance of studying functional computations in their own right, and not simply as a little understood adjunct of set recognition problems, which are the traditional domain of complexity theory.

Of the published work relating functional computations to set recognition problems, perhaps Krentel's work in [Kre88] makes the most explicit use of functional hierarchies. In a manner analogous both to the way one defines the Boolean hierarchy over NP and to the way one defines the set $\Delta^{\mathrm{P}}_2 =_{def}$ $\mathrm{P^{NP}}$ at the second level of the polynomial time hierarchy, for any well-

---

[14] Although he concentrates on set recognition problems, in [Wag88a] Wagner provides a very nice survey covering a wide variety of topics in this area (adaptive versus parallel queries, complete problems for bounded query classes, small numbers of queries versus sparse oracles, polynomial truth-table reducibilities and the Boolean hierarchy).

behaved polynomially computable and polynomially bounded function b, Krentel defines the functional class

$$\text{FP}^{SAT}[b] =_{def} \{f : f \text{ can be computed in polynomial time using at}$$
$$\text{most } b(|x|) \text{ queries to } SAT \text{ for each input } x\}.$$

The union of these classes $\text{FP}^{SAT}[b]$ over all polynomial bounds b gives all of the functions which can be computed in polynomial time given a set in NP as oracle, and is thus directly analogous to the set recognition class Δ_2^P. These functional hierarchies may be of particular interest when the bounding functions, b, are subpolynomial. In this case, the *self-reducibility* of sets like SAT has been well-exploited. As just one example, we next discuss Krentel's work, ([Kre88]), in more detail.

With problems in NP, we traditional associate with accepting paths the simple answer "yes," which gives only limited information about the problem. In the case of a problem like *CLIQUE*, we might want to get more information by associating with any accepting path the *size* of the clique (if any) that is found along the path. This gives more information than a simple "yes" answer, but less information than would be supplied by outputting the clique itself. In the case of a problem like *Traveling Salesperson*, one might want to output the length of the tour found by the path, or even the tour itself. And with SAT we can well imagine wanting each successful path to output the satisfying assignment which it finds. In the terminology used in [Wag88b], these output variations amount to changing the *valuation functions, β_A*, which Wagner assigns to the accepting paths of the validations that $x \in A$.

More interestingly, for a problem like *BIN PACKING* for which there is a polynomial time algorithm which approximates the optimal solution to within $O(log^2(|x|))$, ([KK82]), one might want to output for each path the difference of the path's proposed solution and the calculated polynomial time approximation to the optimal solution. This would enable one to recover the optimal solution from the difference calculated along the path.

Now consider nondeterministic Turing machines which, along each "accepting" computation path produce some functional output value. For any such machine N, define $OptP_N(x)$ to be the largest value produced on any accepting computation path for input x. (Or, for problems like *TSP*, take the smallest value.) Clearly, many problems which we normally associate with nondeterministic set recognition machines can be more crisply formulated as questions about calculating functions in the class, OptP, of functions computed in this manner by nondeterministic machines. For any well-behaved, polynomially computable bounding function b, Krentel defines

$$\text{OptP}[b] =_{def} \{f : f \in \text{OptP} \text{ and } |f(x)| \le b(|x|)\}.$$

Note that *a priori*, b is *not* a bound on any computational aspect of f; it merely bounds the size of the output. Nevertheless, because of the disjunctive self-reducibility of SAT, it is easy to establish a close relationship between the classes OptP[b] and the functional hierarchies FP[b] :

Definition 5.3.1 ([Kre88]) *For functions f and g, define $f \leq^P_{1-tt} g$ if there exist polynomially computable functions, t_1 and t_2 for which $f(x) = t_2(x, g(t_1(x))$.[15] Furthermore, call such a reduction linear if the outer component t_2 can be split into linear components t_3 and t_4. I.e.,*

$$f(x) =_{def} t_2(x, g(t_1(x))) = t_3(x) * g(t_1(x)) + t_4(x).$$

Using the obvious self-reducibility of the functions in OptP, it is not difficult to prove that up to $\leq^P_{1-tt}$ reductions, the classes OptP[b] and $\text{FP}^{SAT}[b]$ are the same for each b, ([Kre88]). (The full power of $\leq^P_{1-tt}$ reductions is not used here.) Furthermore, for functions f in OptP and for linear $\leq^P_{1-tt}$ reductions, there is a tight relation between the functional query hierarchies and more traditional set hierarchies. For example:

Theorem 5.3.2 ([Kre88])
1. If f is complete for OptP under linear reductions, then $\{\langle x, k_1, k_2 \rangle : f(x) = k_1 (mod\ k_2)\}$ is $\leq^P_m$-complete for Δ^P_2.
2. If f is hard for OptP[2] under linear reductions, then $\{\langle x, k \rangle : f(x) = k\}$ is $\leq^P_m$-complete for D^P.
3. If f is hard for OptP[1] under linear reductions, then $\{\langle x, k \rangle : f(x) \geq k\}$ is $\leq^P_m$-complete for NP.

The following are examples of functional problems which Krentel shows to be complete for OptP: *traveling salesperson, lexicographically maximal satisfying assignment, value of knapsack solutions.* For OptP(log), Krentel shows the following functional problems to be complete: *maximal number of satisfying assignments, size of largest clique, the chromatic number of a graph, the longest cycle in a graph, etc.*

Theorems like the preceding can thus be used to project naturally occurring *functional* problems onto set recognition problems which occur above NP in the polynomial time and Boolean hierarchies. We refer the reader to [Kre88] and to [Wag88b] for further details.

Using the Karmarkar-Karp approximation algorithm for *BIN PACKING* discussed earlier, the optimal number of bins can be approximated to within $O(log(log))$ bits, which thus places the problem of calculating the optimal number of bins in OptP($log(log)$). Whether this problem is complete for this class is unknown. However, in view of the following theorem, it cannot be complete for OptP(g) for $g \geq log$ without collapsing NP to P:

---

[15]Krentel calls $\leq^P_{1-tt}$ a *metric* reduction, presumable because it will preserve the number of queries in oracle computations. It is also clearly the correct generalization of *one bounded-truth-table* reductions in the functional context.

Theorem 5.3.3 ([Kre88]) *For honest and monotone functions f and g with $f < g$ and with $f \leq (1 - \epsilon) \cdot log$, if $\mathrm{FP}^{SAT}[f] = \mathrm{FP}^{SAT}[g]$ then* P $=$ NP.

The proof of this theorem is obtained by exploiting the disjunctive self-reducibility of the optimal solution of SAT. Observe that for any Boolean formula $B(x_1, \ldots, x_n)$ the function which on input B gives the first $g(n)$ bits in the maximal satisfying assignment to $x_1, \ldots, x_n$ is a function in $\mathrm{FP}^{SAT}[g]$. Thus it can be calculated by an $\mathrm{FP}^{SAT}[f]$ machine. But since $f < log$, we can simulate the results of the answers to all possible oracle queries in time polynomial in $|x|$ *without actually querying the oracle.*

Simulating the $\mathrm{FP}^{SAT}[f]$ machine allows us to produce a new Boolean formula

$$B'(y_1, \ldots, y_{f(n)}, x_{g(n)+1}, \ldots, x_n, x_1, \ldots, x_{g(n)})$$

such that *any* truth assignment to the variables $y_1, \ldots, y_{f(n)}$ forces a unique assignment to $x_1, \ldots, x_{g(n)}$. This process may be iterated repeatedly until an equivalent formula is obtained which has at most $log(n)$ truly "free" variables. Once this is achieved, we can substitute the n possible satisfying assignments to these $log(n)$ variables, into *this* formula. For each of these possible satisfying assignments, we can calculate the required values for the remaining variables, test the resulting assignment to see if it satisfies the formula, and from these n possible satisfying assignments read off the maximal satisfying assignment. This gives a polynomial time decision procedure for SAT.[16]

Krentel's theorem, which establishes that the $\mathrm{FP}^{SAT}[b]$ query hierarchy is a true functional hierarchy up to logarithmic bounds (assuming that P $\neq$ NP), lends further insight into why problems such as finding the size of the largest clique, which is complete for $\mathrm{FP}^{SAT}[log]$, and finding the length of an optimal traveling salesperson tour, which is complete for FP^{SAT}, should indeed be viewed as having increasing levels of difficulty. And, as pointed out by Krentel, this functional hierarchy makes the problem of characterizing the optimal number of bins for *BIN PACKING*, which lies in $\mathrm{FP}^{SAT}[loglog]$, particularly intriguing.

We should point out that, using similar methods, Wagner, ([Wag88b]), has also obtained many results like those above, and he does so in a nicely systematic fashion. For our purposes, we have concentrated on Krentel's

---

[16]Note that, while this process does not reduce the actual *size* of the formulas, in the larger sense it is still a disjunctive self-reduction since it does produce a formula which is smaller in the partial ordering of formulas obtained by *counting only the free variables.* However, since complexity is measured from the size of the *initial* formula one must worry about the size of the intermediate formulas actually produced. For details concerning this matter see [Kre88].

exposition simply because it makes more explicit use of functional methods and gives a particularly nice result about functional hierarchies.

5.4   Using Self-reducibility to Characterize Polynomial Time

In the preceding sections we discussed results which show how the self-reducibility of SAT can be used to prove that certain problems cannot be solved with only a polynomial amount of information unless P = NP or some other collapse occurs in the polynomial time hierarchy.

In this section, we will consider generalizations of the notion of self-reducibility which can be used to characterize polynomial time. Our starting point is two early results of Selman, ([Sel82a]), which show that P can be characterized as those sets which are both disjunctive self-reducible and p-selective, and also as those p-selective sets which are positive truth-table reducible to their complements.

Definition 5.4.1 ([Sel79]) *A set S is p-selective if there is a polynomially computable function s which, given elements x and y computes $s(x,y) \in \{x,y\}$, and furthermore $s(x,y) \in S$ if and only if at least one of x and y is in S.*

The notion of p-selective comes directly from the classical notion of "semi-recursive" studied by Jockusch in papers in classical recursion theory, ([Joc68], see [Sel82b]).

Thus, from x and y, s is guaranteed to select "a most likely candidate" for membership in S. Ko, ([Ko83]), has characterized the p-selective sets as those sets, S, for which there is a polynomially computable *preordering* for which S is a Dedekind left cut in the induced linear ordering. As a result of this, one can easily show that there are p-selective sets of arbitrarily high computational complexity.

Although p-selective sets are not Turing self-reducible in the sense defined earlier, p-selectivity is a weak form of the *intuitive* notion of self-reducibility since it relates the membership question for any *two elements* to the question of membership for the smaller of the two elements (in some suitable ordering).

Here we are interested in Selman's proof that one can characterize the polynomially decidable sets as those sets which are both disjunctive self-reducible and p-selective. To see this, simply observe that if a set, S, is p-selective, then we can prune the self-reducibility tree as follows: in a linear pass over the daughters of the elements at any node (doing a *breadth first search*) we can find an element "most likely" to be in the set S. Since the self-reduction tree is disjunctive, we can prune all daughters except this most likely daughter from the tree. Continuing in this fashion, in a polynomial number of steps at each level of the self-reduction tree, we can prune the entire self-reduction tree to a "telephone pole." The root node is then in the set if and only if the single leaf of this pruned tree is in S.

Recently, working with Judy Goldsmith, we have used variants of Balcázar's notion of word decreasing query self-reducibility, p-cheatability, and self-reducibility to give new characterizations of the class of sets decidable in polynomial time.

Definition 5.4.2 ([GJY87]) *A set A is near-testable if there is a polynomially computable function which, given $w \neq 0$, decides whether exactly one of w and $w - 1$ is in A, where $w - 1$ is the lexicographic (or numerical) predecessor of w. That is, the function*

$$f(w) = \chi_A(w) + \chi_A(w - 1) \ (mod \ 2)$$

is computable in polynomial time.[17]

For later use notice that if a set is near-testable, then it is word decreasing query (wdq) self-reducible with $w - 1$ being the only query necessary to answer membership for input w. Obviously, all wdq self-reducible sets that require only one query to the set, including all near-testable sets, are decidable in polynomial space. We originally became interested in studying near-testable sets because they are an especially easily defined class of sets which lie someplace between P and PSPACE but require no appeals to "complicated" concepts such as nondeterminism or randomness for their definition. In particular, one can think of them simply as the class of sets whose *boundaries* are polynomially decidable sets.

The other variant of self-reducibility we will be interested in is *p-cheatability*. The *p-cheatable* sets were first studied by Amir, Beigel, Gasarch, Gill, Hay and Owings, ([AG88,Bei87,Bei90,BGGO86,BGH87, BGO87]). The definition of the class is based on the following scenario: one is given a large collection of inputs, $w_1, w_2, ..., w_n$, and one would like to determine whether or not each input is in a known set A. It is assumed that the membership problem for A is difficult. Therefore, if knowing whether some w_i is in A helps in determining whether another w_j is in A, then one would like to use this information. These authors use this idea to motivate the following machine model. To determine membership in A, one designs an oracle machine M that, when given A as oracle will determine membership in A for n different inputs by asking the oracle as few questions

---

[17]Obviously, this definition can be generalized to other exponentially well-founded linear orderings. In [GHJY87] we have provided such a definition.

and doing as little computation as possible. For example, one might hope to design machines that run in polynomial time and for a *fixed* k decide membership for 2^k inputs by asking only k questions. The definition can be formalized as follows.

Definition 5.4.3 ([AG88,Bei87]) *A set A is (n for k) p-cheatable if there exists a polynomial time oracle machine M using A as oracle such that, if M^A is given inputs $(w_1, ..., w_n)$, then, with k or fewer queries to the oracle, M^A determines membership in A for each of $w_1, ..., w_n$. (We will be primarily interested in the case where $n = 2^k$ and in this paper we often call $(2^k$ for $k)$ p-cheatable sets simply p-cheatable.)*

Note that the queries that M asks in the course of deciding membership for $w_1, ..., w_{2^k}$ are *not* restricted to come from the set $\{w_1, ..., w_{2^k}\}$. In fact, for our purposes the exact questions asked and the oracle to which they are addressed turn out not to be relevant.[18] It will turn out that the only thing that matters is the *number* of questions that are asked. Notice that n and k are constants that will depend on the set A.

To put the results that follow in context it is useful to observe the following.

- Amir and Gasarch ([AG88]) have shown that there are p-cheatable sets of arbitrarily high time complexity. Thus, there are p-cheatable sets that are not wdq self-reducible, and thus not near-testable. These results are extended in [GJY90a], where we construct p-cheatable sets that are *bi-immune* with respect to various complexity classes.

- In [GJY87] we have shown that if one-way functions exist, then there are near-testable sets not in P. In fact, in a weak sense, the class of near-testable sets turns out to be $\leq^P_{1-1}$ equivalent to the Papadimitriou-Zachos class, *parity*-P, ([GHJY87]).

Thus it is highly likely that both the near-testable sets and the p-cheatable sets are proper extensions of the class P of sets decidable in polynomial time.

While the analysis of $(2^k$ for $k)$ p-cheatable sets for arbitrarily large (but) fixed k often involves more difficult combinatorial arguments, one frequently, but not always, gains the necessary fundamental insight for analyzing the general case by considering the special case $k = 1$. For example, in the case $k = 1$, if we are given two elements a and b and we consider

---

[18]Beigel's definition of p-cheatable sets (sets which we call $(2^k$ for $k)$ p-cheatable) allows M access to *any* fixed oracle. However, the definitions given by Amir and Gasarch for $(2^k - 1$ for $k)$ p-cheatable sets require that M use A as the oracle. (These sets are called *verbose* by Amir and Gasarch.)

any algorithm which decides both $a \in S$ and $b \in S$ by making only one oracle call, then without consulting the oracle we can ask what the outcomes would be if the oracle returned the answer "yes" *and also* if the oracle returned the answer "no." Now one possibility is that independent of whether the oracle answers "yes" or "no" the algorithm decides, for example, that $a \in S$. In this case, in polynomial time we will have learned that one of the pair, namely a, is in S. What happens if the algorithm does not locate either a or b in S or in $\overline{S}$? Then the algorithm must yield different answers to the question "$a \in S$?" depending on whether the oracle answers "yes" or "no." And in this case the same thing must happen for b. If the patterns for a and for b agree, then we know that $a \in S \iff b \in S$, while if the patterns for a and for b disagree, then we know that $a \in S \iff b \notin S$. Summarizing, if S is (2 for 1) p-cheatable, then there is a polynomial time algorithm which, (without the use of an oracle), given any a and b either places at least one of a or b in S or in $\overline{S}$, or else the algorithm tightly binds the membership questions of a and b.[19]

In the case where a and b are adjacent elements, this relationship is very close to the near-testability relationship. But there is a significant difference. Suppose we knew *a priori* that $a \in S$. Then the near-testability relationship would enable us to decide whether $b \in S$, but if we applied the cheatability relationship then we might simply again get the information that $a \in S$, which by itself is of no help in determining whether $b \in S$.

The next result shows that, in spite of the fact that it is unlikely that either can be used alone to characterize polynomial time, near-testability and p-cheatability *can* be combined to characterize polynomial time.

Theorem 5.4.4 ([GJY90b]) *For any fixed k, a set S is in* P *if and only if S is both near-testable and $(2^k$ for $k)$ p-cheatable.*[20]

If S is in P, the result is of course trivial. For the converse, we prove the result here only for $k = 1$, referring the reader to [GJY90b] for the proof for arbitrary k.

The idea is as follows. Suppose we have $t+1$ integers, $a_0 < a_1 < \ldots < a_t$, and we know whether or not $a_0 \in S$ and furthermore for each k, $(0 < k < t)$, we have a relation which, given the answer to $a_k \in S$ provides the answer to $a_{k+1} \in S$. We call the interval from a_0 to a_1 the *critical region* because if we could determine a relationship which, given the answer to $a_0 \in S$ would determine a_1, we would then know the membership of *all* of the a_i. Obviously, if $a_1 = a_0 + 1$, we could use the *near-testability* of S to determine

---

[19]The result in this paragraph (for $k = 1$) appears in [GJY87] and as Lemma 5.4.9 of [Bei87]. An outline of the corresponding result for arbitrary k appears in [GJY87] and in full detail as part of the proof of Theorem 1 in [GJY90b].

[20]Richard Beigel has independently obtained this theorem for the special case $k = 1$.

such a relationship. This will give us the *final* step of our algorithm. The algorithm for testing membership in S now goes as follows:

To decide whether an element $b \in S$, begin by setting $a_0 = 0$ and $a_1 = a_t = b$. This makes the interval from 0 to b the critical region. Proceeding recursively then, if $a_1 = a_0 + 1$, we use near-testability on a_0 and a_1 and we are done. Otherwise, let c be the midpoint between a_0 and a_1. Apply the p-cheatability algorithm to the pair c and a_1. Three things can happen.

- If the p-cheatability algorithm tells us whether or not $a_1 \in S$, we are done, for we can then successively find the answers to membership for each a_i.

- If the p-cheatability algorithm tells us whether or not $c \in S$, we can reset $a_0 = c$ and try again. In this case we have cut the length of the critical region by half.

- If the p-cheatability algorithm tells us how the membership question for $c \in S$ relates to the membership question for $a_1 \in S$, we can insert c between a_0 and a_1, making the interval from a_0 to c the new critical region. To do this we simply relabel each a_i as a_{i+1} for all $i \geq 1$ and then reset a_1 to c. Note that this again cuts the length of the critical region by half.

Since at each basic iteration of the algorithm, either the algorithm terminates or we cut the size of the critical region by half, the total number of iterations cannot exceed $log(b)$. It is thus easy to see that this method gives a polynomial time algorithm for S. The idea for $k > 1$ is similar, but the combinatorics are more difficult. We refer the reader to [GJY90b] for details.

The question of whether this result can be proved with a word decreasing query condition replacing the near-testability condition is open. Again, some partial results may be found in ([GJY90b]).

In closing, we shall discuss one final characterization of the polynomial time decidable sets using variations of self-reducibilities. We remarked above that for p-cheatable sets, when we reduce the membership question for 2^k elements of the set to k elements of the set then the k elements of the set need not be among the original 2^k elements, or even be of the same or smaller size. Indeed, if given $k + 1$ elements we can always reduce their membership question to k or fewer elements of smaller size, then it is not too difficult to see that given *any* truth-table self-reduction tree (disjunctive or not), we could systematically use this form of strong p-cheatability to force the self-reduction tree to have finitely bounded width. (If at any time any level of the tree had more than k nodes we could use p-cheatability to prune the tree back to k *critical* nodes which are shorter than the original nodes, (although possibly not among the original nodes).)

What is surprising is that this can be made to work for the most general form of self-reducibility and for $(2^k$ for $k)$ p-cheatability *even when the sizes of the queried nodes increase.* This result, was originally proven by Beigel for polynomially *self-truth-table-reducible* sets, ([Bei87, Theorem 5.4.6]. However, the result can also be proved for arbitrary Turing self-reducible sets. Here we merely state the result. We refer the reader to ([GJY90b]) for details of the full proof, or to [Bei87] for details for the simpler case.

Theorem 5.4.5 ([GJY90b]) *For arbitrary k, a set is in* P *if and only if it is both self-reducible and $(2^k$ for $k)$ p-cheatable.*

The question of whether this result can be proved for, say, $(2^k - 1$ for $k)$ p-cheatable is open, although some partial results may be found in ([GJY90b]).

Acknowledgements: The authors would like to thank Eric Allender, Klaus Ambos-Spies, José Balcázar, Lane Hemachandra, and Alan Selman for many useful discussions. Special thanks go to Judy Goldsmith for her contributions to our paper presented at the *Second Annual Structure in Complexity Conference* on which this paper is partially based.

5.5   References

[AG88]   A. Amir and W. Gasarch. Polynomial terse sets. *Inform. and Comput.*, 76, 1988.

[Bal90a]   J. Balcázar. Self-reducibility. *J. Comput. Systems Sci.*, 1990. To appear.

[Bal90b]   J. Balcázar. Self-reducibility structures and solutions of NP problems. *Revista Matematica*, 1990. To appear.

[BBS86]   J. Balcázar, R. Book, and U. Schöning. The polynomial-time hierarchy and sparse oracles. *J. Assoc. Comput. Machinery*, 33(3):603–617, 1986.

[BD76]   A. Borodin and A. Demers. *Some comments on functional self-reducibility and the NP hierarchy.* Technical Report TR76-284, Cornell University, Department of Computer Science, 1976.

[Bei87]   R. Beigel. *Query-limited Reducibilities.* PhD thesis, Stanford University, October 1987.

[Bei90]   R. Beigel. A structural theorem that depends quantitatively on the complexity of SAT. *J. Comput. Systems Sci.*, 1990. This paper was presented at the *2nd Annual Structure in Complexity Theory Conf.* in 1987.

[Ber78]   P. Berman. Relationships between density and deterministic complexity of NP-complete languages. In *5th EATCS Conference in Automata, Languages and Programming*, pages 63–71, Springer-Verlag, Lecture Notes in Computer Science, Berlin, 1978.

[BGGO86] R. Beigel, W. Gasarch, J. Gill, and J. Owings. Terse, superterse and verbose sets. *Preprint*, 1–25, 1986.

[BGH87]   R. Beigel, W. Gasarch, and L. Hay. *Bounded Query Classes and the Difference Hierarchy.* Technical Report 1847, University of Maryland, 1987.

[BGO87]   R. Beigel, W. Gasarch, and J. Owings. Terse sets and verbose sets. *Recursive Function Theory: Newsletter*, 36:13–14, 1987.

[BHZ87]   R. Boppana, J. Håstad, and S. Zachos. Does coNP have short interactive proofs? *Inform. Process. Letters*, 25:127–132, 1987.

[BJY90]   D. Bruschi, D. Joseph, and P. Young. A structural overview of NP optimization problems. In *Proc. of the 2nd International Conference on Optimal Algorithms*, Springer-Verlag, Lecture Notes in Computer Science, 1990. 27 pages.

[BK87]   R. Book and K. Ko. On sets reducible to sparse sets. In *Proc. 3rd Annual Structure in Complexity Theory Conf.*, pages 147–154, 1987.

[BS90]   J. Balcázar and U. Schöning.   Logarithmic advice classes. *EATCS J. Theor. Comput. Sci.*, ():, 1990. To appear.

[For79]   S. Fortune. A note on sparse complete sets. *SIAM J. Comput.*, 8:431–433, 1979.

[GHJY87] J. Goldsmith, L. Hemachandra, D. Joseph, and P. Young. *Near-testable sets.* Technical Report , University of Wisconsin, Computer Sciences Dept., 1987.

[GJY87]   J. Goldsmith, D. Joseph, and P. Young. Self-reducible, p-selective, near-testable, and p-cheatable sets: the effect of internal structure on the complexity of a set. In *Proc. 2nd Annual Structure in Complexity Theory Conf.*, pages 50–59, 1987.

[GJY90a]   J. Goldsmith, D. Joseph, and P. Young. *A note on bi-immunity and p-closeness of p-cheatable sets in P poly.* Technical Report 741-1987, University of Wisconsin, Computer Sciences Dept., 1990. [To appear in *J. Comput. Systems Sci.*].

[GJY90b]   J. Goldsmith, D. Joseph, and P. Young. *Using self-reducibilities to characterize polynomial time.* Technical Report 749-1987, University of Wisconsin, Computer Sciences Dept., 1990. [To appear in *Inform. and Comput.*].

[HH87]   J. Hartmanis and L. Hemachandra. One-way functions, robustness, and the non-isomorphism of NP-complete sets. In *Proc. 2nd Annual Structure in Complexity Theory Conf.*, pages 160–174, 1987. To appear in the *EATCS J. Theor. Comput. Sci.*

[Imm88]   N. Immerman. Nondeterministic space is closed under complementation. *SIAM J. Comput.*, 17(5):935–938, 1988.

[JLY89]   D. Joseph, T. Long, and P. Young. How bounded truth-table reductions to sparse sets collapse the polynomial time hierarchy. *Preprint*, 10, 1989.

[Joc68]   C. Jockusch. Semirecursive sets and positive reducibility. *Trans. Amer. Math. Soc.*, 131:420–436, 1968.

[Kad87]   J. Kadin. $P^{NP[\log n]}$ and sparse Turing-complete sets for NP. In *Proc. 1st Annual Structure in Complexity Theory Conf.*, pages 33–40, 1987.

[KK82]   N. Karmarkar and R. Karp. An efficient approximation scheme for the one dimensional bin-packing problem. In *Proc. 23rd IEEE Symp. Found. Comput. Sci.*, pages 312–320, 1982.

[KL80]   R. Karp and R. Lipton. Some connections between nonuniform and uniform complexity classes. In *Proc. 12th ACM Symp. Theory of Comput.*, pages 302–309, 1980. [Also published as, Turing machines that take advice, *L'Ensignement Mathématique*, 28:191-210, 1982.].

[Ko83]   K. Ko. On self-reducibility and weak P-selectivity. *J. Comput. Systems Sci.*, 26:209–211, 1983.

[Ko87]   K. Ko. On helping by robust oracle machines. In *Proc. 2nd Annual Structure in Complexity Theory Conf.*, pages 182–190, 1987.

[Ko88]   K. Ko. Distinguishing bounded reducibilities by sparse sets. In *Proc. 3rd Annual Structure in Complexity Theory Conf.*, pages 181–191, 1988.

[Kre88]   M. Krentel. The complexity of optimization problems. *J. Comput. Systems Sci.*, 36:490–509, 1988.

[Lon82]   T. Long. Strong nondeterministic polynomial-time reducibilities. *EATCS J. Theor. Comput. Sci.*, 21:1–25, 1982.

[Mah82]   S. Mahaney. Sparse complete sets for NP: solution of a conjecture of Berman and Hartmanis. *J. Comput. Systems Sci.*, 25:130–143, 1982.

[Mah86]   S. Mahaney. Sparse sets and reducibilities. In R. Book, editor, *Studies in Complexity Theory*, pages 63–118, Pitman, 1986.

[Mah89]   S. Mahaney. The isomorphism conjecture and sparse sets. In J. Hartmanis, editor, *Computational Complexity Theory*, page , American Mathematical Society, 1989.

[MP79]   A. Meyer and M. Paterson. *With What Frequency are apparently Intractable Problems Difficult?* Technical Report MIT/LCS/TM-126, M.I.T., 1979.

[PY84]   C. Papadimitriou and M. Yannakakis. The complexity of facets and some facets of complexity. *J. Comput. Systems Sci.*, 28:244–259, 1984.

[PY88]   C. Papadimitriou and M. Yannakakis. Optimization, approximation and complexity classses. In *Proc. 20th ACM Symp. Theory of Comput.*, pages 229–234, 1988.

[Sch76]   C. Schnorr. Optimal algorithms for self-reducible problems. In *3rd EATCS Conference in Automata, Languages and Programming*, pages 322–337, Edinburgh, University Press, July 1976.

[Sch79]   C. Schnorr. On self-transformable combinatorial problems. *Symp. on Math. Optimierung, Oberwolfach*, 1979.

[Sch85]   U. Schöning. Robust algorithms a different approach to oracles. *EATCS J. Theor. Comput. Sci.*, 40:57–66, 1985.

[Sel79]   A. Selman. P-selective sets, tally languages, and the behavior of polynomial time reducibilities on NP. *Math. Systems Theory*, 13:55–65, 1979.

[Sel82a]   A. Selman. Analogues of semirecursive sets and effective re-
            ducibilities to the study of NP complexity. *Inform. and Con-
            trol*, 52(1):36–51, 1982.

[Sel82b]   A. Selman. Reductions on NP and P-selective sets. *EATCS J.
            Theor. Comput. Sci.*, 19:287–304, 1982.

[Sel84]   A. Selman. *Remarks About Natural Self-Reducible Sets in NP
            and Complexity Measures for Public-key Cryptosystems*. Tech-
            nical Report, Iowa State University, 1984.

[Sel88a]   A. Selman. Natural self-reducible sets. *SIAM J. Comput.*,
            17:989–996, 1988.

[Sel88b]   A. Selman. Promise problems complete for complexity classes.
            Inform. and Comput., 78:87–98, 1988.

[Sze88]   R. Szelepcsényi. The method of forced enumeration for nonde-
            terminisitc automata. *Acta Inform.*, 26:279–284, 1988.

[Tra70]   B. Trachtenbrot. On autoreducibility. *Dokl. Akad. Nauk.
            SSSR*, 11:335–350, 1970.

[Ukk83]   E. Ukkonen. Two results on polynomial time Turing reductions
            to sparse sets. *SIAM J. Comput.*, 12:580–587, 1983.

[Wag88a]   K. Wagner. Bounded query computations. In *Proc. 3rd Annual
            Structure in Complexity Theory Conf.*, pages 260–277, 1988.

[Wag88b]   K. Wagner. More complicated questions about maxima and
            minima, and some closures of NP. *EATCS J. Theor. Comput.
            Sci.*, 51:53–80, 1988.

[WW86]   K. Wagner and G. Wechsung. *Computational Complexity*. D.
            Reidel Publishing Co., 1986.

[Yap83]   C. K. Yap. Some consequences of non-uniform conditions
            nonuniform classes. *EATCS J. Theor. Comput. Sci.*, 26:287–
            300, 1983.

[Yes83]   Y. Yesha. On certain polynomial-time truth-table reducibilities
            of complete sets to sparse sets. *SIAM J. Comput.*, 12:411–425,
            1983.

6

The Structure of Complete Degrees

Stuart A. Kurtz[1]
Stephen R. Mahaney[2]
James S. Royer[3]

6.1 Introduction

The notion of NP-completeness has cut across many fields and has provided a means of identifying deep and unexpected commonalities. Problems from areas as diverse as combinatorics, logic, and operations research turn out to be NP-complete and thus computationally equivalent in the sense discussed in the next paragraph. PSPACE-completeness, NEXP-completeness, and completeness for other complexity classes have likewise been used to show commonalities in a variety of other problems. This paper surveys investigations into how strong these commonalities are. More concretely, we are concerned with:

> What do NP-complete sets look like?
>
> To what extent are the properties of particular NP-complete sets, e.g., SAT, shared by *all* NP-complete sets?
>
> If there are are structural differences between NP-complete sets, what are they and what explains the differences?

We make these questions, and the analogous questions for other complexity classes, more precise below. We need first to formalize NP-completeness.

There are a number of competing definitions of NP-completeness. (See [Har78a, p. 7] for a discussion.) The most common, and the one we use, is based on the notion of *m-reduction*, also known as *polynomial-time many-one reduction* and *Karp reduction*. A set A is *m-reducible* to B if and only if there is a (total) polynomial-time computable function f such that

$$\text{for all } x, \quad x \in A \iff f(x) \in B. \tag{1}$$

---

[1]Department of Computer Science; University of Chicago; 1100 E. 58th St; Chicago, IL 60637.

[2]Department of Computer Science, University of Arizona, Tucson, AZ 85721.

[3]School of Computer & Information Science; Syracuse University; Syracuse, NY 13210.

Intuitively, (1) says that A is no harder than B in the sense that, if one has a means of answering "$x \in B$?" questions, then one can also answer "$x \in A$?" questions with the computation of $f(x)$ as the the only additional overhead. An NP-*complete set* is an NP set to which all NP sets are m-reducible. Thus, any two NP-complete sets are m-equivalent, that is, interreducible by m-reductions. Since m-reductions relate the hardness of sets, the NP-complete sets are therefore all of the same degree of difficulty.

The formal definition of m-reduction does not seem to capture all the important aspects of the reductions used by Karp and his successors in NP-completeness proofs. In a Karp-style proof that a problem B is NP-complete, one takes a known NP-complete problem A, and shows how to *translate* the structure of an arbitrary instance of A into an instance of B. (See [GJ79, Chapter 3] for examples of such arguments.) Since these "in practice" m-reductions preserve the structure of individual instances, one might expect that these reductions would also preserve the global structure of a problem, and, that the NP-complete sets shown equivalent by Karp-style proofs would share a strong structural similarity.

Berman and Hartmanis [BH77] showed that this is the case in 1977. They considered a number of well known NP-complete sets and showed that these sets are all pairwise *polynomial-time isomorphic*.[4] Since many complexity theoretic properties of sets are invariant under polynomial-time isomorphisms, Berman and Hartmanis thus established that, in many respects, these NP-complete sets are *identical* in structure. Moreover, Berman and Hartmanis went through the current literature on "conventional" NP-complete sets[5] and observed that using their methods these sets could also be shown to be polynomial-time isomorphic to their previous examples. Motivated by this evidence, the lack of any counterexamples, and also by analogy with results in recursion theory, Berman and Hartmanis made the following conjecture.

The Isomorphism Conjecture *All* NP-*complete sets are polynomial-time isomorphic.*

This conjecture predicts that $P \neq NP$, that NP-complete sets are paddable (see S 6.2) and self-reducible (see, for example, [Mah89]), and that NP-complete sets and their complements have exponential density.[6] This

---

[4]A is *polynomial-time isomorphic to* B if and only if there exists f, an m-reduction from A to B as in (1), that is also 1-1, onto, and whose inverse f^{-1} is computable in polynomial-time, in other words, f is a polynomial-time computable and invertible permutation on $\{0,1\}^*$ such that $f(A) = B$ and $f(\overline{A}) = \overline{B}$.

[5]By a *conventional* NP-complete set we mean, informally, a set that results from a straightforward coding of an NP-complete problem that stems from work in combinatorics, optimization, etc.

[6]A has *exponential density* if and only if the growth of Cardinality($\{x \in A : |x| \leq n\}$) is $\Omega(2^{n^{1/k}})$ for some k.

last prediction was partially confirmed by P. Berman [Ber78] and Fortune [For79] who showed that if P $\neq$ NP, then there are no coNP-complete sparse sets[7] and by Mahaney [Mah82] who showed that if P $\neq$ NP, then there are no NP-complete sparse sets.[8]

The isomorphism conjecture is attractive and intriguing, but there are good reasons to doubt the sufficiency of its supporting evidence. Let us call the sets p-isomorphic to SAT *standard* NP-complete sets.[9] The NP-complete sets that Berman and Hartmanis showed to be standard are all conventional NP-complete sets as discussed above, and, moreover, since Berman and Hartmanis's sets are all p-isomorphic, they each can be rightly viewed as being a natural encoding of a *single* combinatorial problem— Satisfiability. However, there is no *a priori* reason to expect that every NP-complete set is conventional, much less standard. Joseph and Young [JY85] raised this objection and backed it up by constructing nonconventional NP-complete sets that are not obviously standard. Moreover, they conjectured that some of their NP-complete sets do indeed fail to be standard.

One-way functions[10] played a key role in Joseph and Young's examples. Selman has shown that the Joseph-Young sets have the form $f(A)$ for some standard NP-complete set A and some one-way function f. (Watanabe [Wat85] made related observations.) One can think of $f(A)$ as a highly encrypted version of A. Thus, the essential idea behind Joseph and Young's conjecture is

The Encrypted Complete Set Conjecture   *There is a one-way function f and a standard complete set A such that A and the NP-complete set $f(A)$ are not polynomial-time isomorphic.*

One problem with this conjecture is that there is no concrete candidate for the one-way function of the conjecture. Furthermore, any one-way function that fits the conjecture's requirements would seemingly need to possess much stronger properties than just the failure to have a polynomial-time inverse. What these properties should be is uncertain.

The contention between the isomorphism and the encrypted complete set conjectures has provoked much study. Other than the sparse set re-

---

[7]A set A is *sparse* if and only if the growth of Cardinality($\{x \in A : |x| \leq n\}$) is $\mathcal{O}(n^k)$ for some k.

[8]This paper does not further discuss sparse sets results. Mahaney surveys this work in [Mah89].

[9]By Theorem 6.4.6 below and the fact that paddability (defined in S 6.2) is preserved under p-isomorphisms, the standard NP-complete sets are precisely the paddable NP-complete sets.

[10]A polynomial-time computable function f is *one-way* if and only if f is 1-1 and polynomially honest, but f^{-1} is not polynomial-time computable. These functions are widely conjectured to exist and various forms of one-way functions play important roles in cryptography.

sults mentioned above, the structure of the NP-complete sets remains a (complete) mystery. However, for certain other complexity classes, a reasonable understanding of their m-complete sets[11] has emerged. The best understood class of m-complete sets are those of EXP, the sets decidable in deterministic exponential time.[12] For example, it is known that:

- The EXP-complete sets are all 1-li-equivalent, that is, they are pairwise m-equivalent as witnessed by 1-1, length increasing m-reductions. A 1-li-equivalence is intuitively a very strong equivalence relation on sets. One can show that if one-way functions do not exist, then 1-li-equivalence implies polynomial-time isomorphism. Thus, if one-way functions do not exist, all the EXP-complete sets are polynomial-time isomorphic. (See Theorems 6.4.2 and 6.5.2 below.)

- There is a characterization of the existence of one-way functions involving the 1-li-equivalence of sets: One-way functions exist if and only if there are two sets in EXP that are 1-li-equivalent, but *not* polynomial-time isomorphic. Moreover, one can take these two sets to be "close" to m-complete for EXP. (See Theorem 6.6.2 below.)

- The isomorphism conjecture asserts that a particular m-equivalence class of sets (the class of NP-complete sets) "collapses" in the sense that any two members of this m-equivalence class are polynomial-time isomorphic. One can exhibit such a "collapsing" m-equivalence class of sets in EXP. Moreover, the sets in this m-equivalence class are all "close" to m-complete for EXP in the same sense of "close" as the two sets in the previous item. (See Theorem 6.6.1 below.)

- There are certain very strong forms of one-way functions the existence of which would imply that there are "encrypted complete sets" for EXP (and also NP, PSPACE, and many other complexity classes). Such one-way functions do exist relative to random oracles. Therefore, the NP, PSPACE, EXP, ... versions of the encrypted complete set conjecture are correct relative to a random oracle. (See Proposition 6.5.10 and Theorem 6.7.5 below.)

These results illustrate the miles we have come in answering the questions raised by the two conjectures and the parsecs to go until any final resolution is reached. These questions seem far more complex than anyone had initially believed and they seem to have interesting connections to other

---

[11] That is, the sets in a complexity class to which all sets in the class polynomial-time m-reduce.

[12] In this paper, we say that a set is decidable in exponential time if and only if the set has a deterministic decision procedure that, for some k, runs in $2^{O(n^k)}$ time.

fundamental questions in complexity theory, e.g., the nature of one-way functions.

This article discusses research on the isomorphism and the encrypted complete set conjectures. We have *not* attempted to write a comprehensive survey of this work. We have instead tried to present a coherent sketch of the area's key results and ideas. Towards this end we have included proofs that illustrate some of these ideas. Our goal has been to lead the reader from a basic understanding of NP-completeness to the current frontiers of research on the structure of classes of complete sets. Readers will also benefit from Young's survey in this volume, as well as from consulting the original references.

6.2   Basic Definitions

We shall assume that the reader is familiar with the rudiments of recursion theory, e.g., the s-m-n theorem. See [Cut80] and [DW83] for elementary introductions. We also shall assume the reader is familiar with the basics of machine based computational complexity as discussed in [HU79] and with the definitions of LOGSPACE, P, NP, and PSPACE. Σ_k^P denotes the k-th Σ class in the Meyer and Stockmeyer [MS72] polynomial hierarchy. ($\Sigma_1^P = $ NP and $\Sigma_2^P = $ the sets in NP relative to an oracle for SAT. See [GJ79] for more discussion.) EXP denotes the class of sets decidable within a $2^{\text{poly}(n)}$ time bound on a deterministic Turing machine, and NEXP denotes the corresponding nondeterministic time class. RE denotes the class of recursively enumerable sets. Logspace, Ptime, and Pspace respectively denote the class of total functions computable in deterministic logarithmic-space, polynomial-time, and polynomial-space.

N denotes the set of natural numbers $\{0, 1, 2, \ldots\}$. We identify each $x \in N$ with the x-th string over the symbols $\mathbf{0}$ and $\mathbf{1}$ in the lexicographic ordering on $\{\mathbf{0}, \mathbf{1}\}^*$. We tend to use natural numbers and strings over $\{\mathbf{0}, \mathbf{1}\}$ interchangeably. Unless specified otherwise, functions are over N and total and sets are subsets of N. The length of $x \in N$ (i.e., the length of its string representation) is denoted $|x|$. Let $\langle \cdot, \cdot \rangle$ denote a polynomial-time computable pairing function, see [Rog67] for an example.

CONDITIONS ON FUNCTIONS. A function f is:

h-honest if and only if, for all x, $|x| \leq h(|f(x)|)$.

polynomially honest if and only if for some polynomial p, f is p-honest.

exponentially honest if and only if for some constant c, f is $\lambda n. 2^{cn+c}$-honest.

length increasing if and only if, for all x, we have $|f(x)| > |x|$.

linear length bounded if and only if there is a constant a such that, for all x, $|f(x)| \leq a(|x| + 1)$.

polynomial length bounded if and only if there is a polynomial p, such that, for all x, $|f(x)| \leq p(|x|)$.

p-invertible if and only if there is a Ptime function g such that $g \circ f$ is the identity.

REDUCIBILITIES. Recall from the introduction that a set A is *m-reducible* (i.e., many-one reducible) to a set B if and only if there is a polynomial-time computable function f such that

$$\text{for all } x \in N, \quad x \in A \iff f(x) \in B. \tag{2}$$

A is *recursively m-reducible to B* if and only if there is a recursive function f such that (2) holds. Logspace, Pspace, ... m-reductions are defined analogously.

A is *1-reducible* to B if and only if there is a 1-1 function which witnesses that A is m-reducible to B. A is *1-honest-reducible* to B if and only if there is a polynomially honest function which witnesses that A is 1-reducible to B. A is *1-li-reducible* to B if and only if there is a length increasing function which witnesses that A is 1-reducible to B or else $A = B$. (The $A = B$ clause is to make 1-li-reducibility a reflexive relation.) Finally, A is *p-isomorphic* to B if and only if there is an f which witnesses that A is m-reducible to B which is also 1-1, onto, and p-invertible. **N.B.** If A and B are p-isomorphic, then they are necessarily 1-honest-equivalent, but they need not be 1-li comparable.[13]

A set A is *complete* for a class of sets $\mathcal{C}$ with respect to a reducibility R if and only if A is in $\mathcal{C}$ and every set in $\mathcal{C}$ R-reduces to A. So, for example, we can speak of m-complete, 1-complete, 1-li-complete, ... sets for NP. In this paper "$\mathcal{C}$-complete" means "m-complete for the class $\mathcal{C}$."

Reducibilities relate the hardness of sets. Hence, an equivalence class of sets with respect to a reducibility relation consists of sets of the same "degree" of difficulty. We thus define a *degree* to be an equivalence class under a reducibility.[14] So, we speak of m-degrees, 1-degrees, 1-honest-degrees and 1-li-degrees according to the reducibility intended.

A degree *collapses* if and only if all of its elements are pairwise p-isomorphic. Thus, the Berman-Hartmanis isomorphism conjecture can be stated succinctly: the complete m-degree for NP collapses. We sometimes use the term "collapses" more generally. For example, we say an m-degree collapses to a 1-li-degree when all of the m-degree's elements are 1-li-equivalent.

CYLINDERS. A set A is a *cylinder* if and only if for some B, A is p-isomorphic to the set $\{\langle b, z \rangle : b \in B \ \& \ z \in N\}$. A set A is *paddable* if

---

[13]Consider the sets $\{0^{4^i}, 0^{2 \cdot 4^i + 1} : i \in N\}$ and $\{0^{4^i + 1}, 0^{2 \cdot 4^i} : i \in N\}$.

[14]The term "degree" comes from Post's [Pos44]—the paper that founded modern recursion theory.

and only if there exists $p(\cdot, \cdot)$, a polynomial-time computable, p-invertible[15] function, such that, for all a and x,

$$a \in A \iff p(a, x) \in A.$$

It is clear that cylinders are paddable. Mahaney and Young [MY85] show that all paddable sets are cylinders. Below we shall treat being a cylinder and being paddable as synonymous. Since paddability is, on the surface, a looser condition on a set, it turns out to be the easier of the two notions with which to work.

PROGRAMMING SYSTEMS. We say $\nu(\cdot, \cdot)$ is a *universal function* for a class of partial functions $\mathcal{F}$ if and only if $\nu(\cdot, \cdot)$ is partial recursive and $\mathcal{F} = \{\lambda x . \nu(i, x) : i \in N\}$. Such a $\nu(\cdot, \cdot)$ determines a *programming system* $\langle \nu_i \rangle_{i \in N}$ for $\mathcal{F}$, where for each i, $\nu_i = \lambda x . \nu(i, x)$. A universal function $\nu(\cdot, \cdot)$ and its associated programming system $\langle \nu_i \rangle_{i \in N}$ are really two different views of the same object. We use ν to denote both the universal function $\nu(\cdot, \cdot)$ and programming system $\langle \nu_i \rangle_{i \in N}$.

A programming system, φ, for the class of partial recursive functions is *acceptable* if and only if for each programming system ψ (for some class of functions), there is a recursive function t such that, for all p, $\psi_p = \varphi_{t(p)}$, i.e., there is an effective way of translating ψ-programs into equivalent φ-programs. (For alternative definitions see [MY78] and [Rog67].) Henceforth, φ will be a fixed acceptable programming system.

We say that $\langle S_i \rangle_{i \in N}$ is a *programming system for a class of sets* S if and only if (i) $[x \in S_i]$ is partial recursive relation in i and x and (ii) $S = \{S_i \mid i \in N\}$. For example, for all i, define $W_i = \text{domain}(\varphi_i)$; $\langle W_i \rangle_{i \in N}$ is then a programming system for the class of r.e. sets [Rog67].

The *recursion theorem* holds for a programming system ν if and only if for each "ν-program" i, there is another ν-program e such that

$$\nu_e = \lambda x . \nu_i(\langle e, x \rangle). \tag{3}$$

Intuitively, e is a self-referential program that, on input x, generates a copy of its own program "text" and subsequently uses that copy as an additional datum together with x on which to run program i.[16] If $\langle A_i \rangle_{i \in N}$ is a programming system for a class of sets, then the recursion theorem holds for $\langle A_i \rangle_{i \in N}$ if and only if, for each i, there is an e such that

$$A_e = \{x : \langle e, x \rangle \in A_i\}.$$

---

[15]In this context, p-invertible means that there is a polynomial time computable g such that, for all x and y, $g(p(x, y)) = \langle x, y \rangle$.

[16]**N.B.** This is the Kleene form of the recursion theorem [Rog67, p. 214], *not* the commonly taught fixed point form. In acceptable programming systems the two forms are roughly equivalent. However, for subrecursive programming systems for natural subrecursive classes, Kleene's form of the recursion theorem is often valid, but the fixed point form is essentially *never* valid. (Consider the possible "fixed points" for an f such that, for all i and x, $\nu_{f(i)} = 1 + \nu_i(x)$.)

The recursion theorem holds for φ [Rog67, p. 214] and also $\langle W_i \rangle_{i \in N}$. Let ψ be a programming system for the polynomial-time computable functions such that $\lambda i, x . \psi_i(x)$ is computable in time exponential in $|i| + |x|$ and such that the recursion holds for ψ. See [Koz80] and [RC87] for examples of such ψ.

Suppose C is an m-complete set for a complexity class $\mathcal{C}$. Observe that f m-reduces a set A to C if and only if $A = f^{-1}(C)$. For all i, let $A_i = \psi_i^{-1}(C)$. It follows that that $\langle A_i \rangle_{i \in N}$ is a programming system for $\mathcal{C}$. Moreover, the recursion theorem holds for $\langle A_i \rangle_{i \in N}$.[17]

For more about programming systems for subrecursive collections of functions and sets, their properties, and their uses, see [RC87].

ONE-WAY FUNCTIONS. A function f is *one-way* if and only if f is 1-1, polynomially honest, and polynomial-time computable, yet not p-invertible. A set A is in the class UP if and only if there exists a polynomial p and a polynomial-time decidable predicate $Q(\cdot, \cdot)$ such that

$$A = \{x : (\exists y : |y| \leq p(|x|)) \, Q(x, y)\},$$

and, for each x, there is at most one y such that $Q(x, y)$. UP is clearly a subclass of NP. Independently, Berman [Ber77], Grollmann and Selman [GS84] [GS88] and Ko [Ko85] observed that one-way functions exist if and only if P $\neq$ UP.[18] Below this equivalence is taken for granted.

A special sort of one-way function is introduced in [KMR89]. A function f is a *scrambling function* if and only if f is 1-1, polynomially honest, and polynomial-time computable and, for all nonempty A, $f(A)$ is not paddable.

6.3 Recursion Theoretic Results

Before discussing polynomial-time complete m-degrees, we first consider some important related results from recursion theory on the structure of

---

[17]Here is a proof sketch of our claim that the recursion holds for $\langle A_i \rangle_{i \in N}$. For convenience, we identify sets with their characteristic functions. Thus, the equation $A_i = \psi_i^{-1}(C)$ can be rewritten $A_i = C \circ \psi_i$. Fix i. By the recursion theorem for ψ there is an e such that $\psi_e = \lambda x . \psi_i(\langle e, x \rangle)$. Hence,

$$A_e = C \circ \psi_e = \lambda x . C \circ \psi_i(\langle e, x \rangle) = \lambda x . A_i(\langle e, x \rangle).$$

Therefore, the claim follows.

[18]Berman did not explicitly consider one-way functions. Rather, he distinguished between the polynomial-time computable and invertible functions (which he called the "P-E" functions), and polynomial-time computable, 1-1, honest functions (which he called "P-1B" functions). From these definitions, he proved that P-E = P-1B if and only if P = UP.

the class of recursively m-complete r.e. sets. The problem of describing this structure was settled by Myhill. He showed

Theorem 6.3.1 ([Myh55]) *Two sets are recursively 1-equivalent if and only if they are recursively isomorphic.*

Theorem 6.3.2 ([Myh55]) *The recursively m-complete r.e. sets are pairwise recursively 1-equivalent.*

From these two theorems one immediately obtains

Corollary 6.3.3 *The recursively m-complete r.e. sets are pairwise recursively isomorphic.*

Corollary 6.3.3 completely describes the recursion theoretic structure of the class of recursively m-complete r.e. sets—there is essentially only one set in the class. These theorems were part of the inspiration for the original paper of Berman and Hartmanis and remain an important influence on the work on polynomial-time m-complete degrees.

We mention a slightly different route to Corollary 6.3.3 which uses the notion of recursive cylinder.

Theorem 6.3.4 *Two recursive cylinders are recursively m-equivalent if and only if they are recursively isomorphic.*

Theorem 6.3.5 *Every recursively m-complete r.e. set is a recursive cylinder.*

Corollary 6.3.3 follows immediately from these last two theorems. Both of these theorems are implicit in Myhill's paper [Myh59], also see [You66].

The proofs of Theorems 6.3.1 and 6.3.2 introduce a number of ideas important for the complexity theoretic work of the following sections. Because of this importance, we sketch proofs for both theorems. Our proof of Theorem 6.3.1 is largely based on Myhill's original proof which in turn is partly based on the standard construction for the Cantor-Bernstein theorem (sometimes known as the Schröder-Bernstein theorem or the Cantor-Schröder-Bernstein theorem) from set theory.[19] Our argument for Theorem 6.3.2 is based on a proof of a related result about acceptable programming

---

[19]Cantor, in his theory of cardinality of sets [Can55], makes the following two basic definitions. Two sets A and B are said to *have the same cardinality* if and only if there is a 1-1 correspondence between A and B; and A *has cardinality less than or equal to that of* B if and only if there is a 1-1 map from A into B, (i.e., A has the same cardinality as a subset of B). In order for these two definitions to make sense, it should be the case that: If A has cardinality less than or equal to that of B *and* B has cardinality less than or equal to that of A, then A and B have the same cardinality. This *is* the case as shown by

systems [Sch75, Theorem 2], which Schnorr credits to an anonymous referee.

Proof Sketch of Theorem 6.3.1 In the following let N' denote a disjoint copy of N. If x is a member of N, then x' denotes the corresponding element of N'. Now, suppose that $A \subseteq N$, $B \subseteq N'$, $f: N \to N'$ recursively 1-reduces A to B, and $g: N' \to N$ recursively 1-reduces B to A. We introduce the

---

The Cantor-Bernstein Theorem *If there is a 1-1 map from A into B and a 1-1 map from B into A, then there is a 1-1 correspondence between A and B.*

The standard proof of this theorem (see the proof of Theorem 6.3.1) is so simple and direct it is difficult to realize that a satisfactory proof of the theorem was hard to obtain. What follows is a partial history of the result taken from Moore's excellent book [Moo82].

1882 Cantor claims (without proof) the theorem in a paper.

1883 Cantor proves the theorem for subsets of $\mathbf{R}^n$ using the Continuum Hypothesis.

1883 Cantor poses the general theorem as an open problem in a letter to Dedekind.

1884 Cantor again claims (without proof) the theorem in a paper.

1887 Dedekind proves the theorem in his notebook and promptly forgets about solving the problem. This particular solution is first published in 1932 in Dedekind's collected works.

1895 Cantor states the theorem as an open problem in a paper.

1896 Burali-Forti proves the theorem for countable sets.

1897 Bernstein (a Cantor student) proves the general theorem using the Denumerability Assumption, a weak form of the axiom of choice. Cantor shows the proof to Borel.

1898 Schröder publishes a "proof" of the theorem. Korselt points out to Schröder that the proof is wrong. Schröder's proof turns out to be unfixable.

1898 Borel publishes Bernstein's proof in an appendix of Borel's 1898 book on set theory and complex functions.

1899 Dedekind sends Cantor an elementary proof of the theorem, which from Moore's description sounds like the proof used today.

1901 Bernstein's thesis appears (with his proof).

1902 Korselt sends in a proof of the theorem (along the lines of Dedekind's proof) to *Mathematische Annalen*. Korselt's paper appears in 1911!

1907 Jourdain points out that the use of the Denumerability Assumption in Bernstein's proof is removable.

Moore suspects that Cantor had a proof of the theorem that used the well ordering principle (which turns out to be equivalent to the axiom of choice). Cantor was unclear for many years whether the well ordering principle was "a law of thought" or something that needed proof. So, the fact that the theorem came and went several times might have been in part a function of Cantor's uncertainty about well ordering.

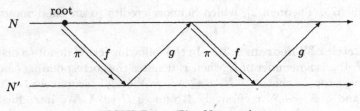

FIGURE 6.1. The N rooted case.

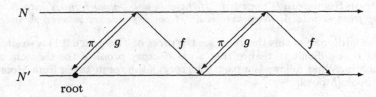

FIGURE 6.2. The N' rooted case.

directed graph $G = (N \cup N', E)$, where

$$E = \{(x, f(x)) : x \in N\} \cup \{(x', g(x')) : x' \in N'\}.$$

G is clearly bipartite. Since f and g are functions, every vertex of G has out-degree one. Since f and g are 1-1, every vertex of G has in-degree of at most one.

The maximal connected components of G we call f, g-*chains*. An f, g-chain C has one of four possible structures.

1. The cyclic case. C is a finite cyclic path containing an even number of vertices.

2. The two-way infinite case. C is an infinite path in which every vertex has in-degree one.

3. The N rooted case. C is an infinite path with a root $r \in N$, i.e., r is a vertex of C with in-degree zero.

4. The N' rooted case. C is an infinite path with a root in N'.

We say a function $h: N \to N'$ *respects* f, g-*chains* if and only if for all x, $h(x)$ is a member of x's f, g-chain. Since f and g recursively 1-reduce A to B and B to A, respectively, it follows that for any h that respects f, g-chains we have, for each x, that $x \in A \iff h(x) \in B$.

Now, given f and g, the standard construction for Cantor-Bernstein defines

$$\pi = \lambda x. \begin{cases} f(x), & \text{if } x\text{'s } f, g\text{-chain is either cyclic,} \\ & \quad \text{two-way infinite, or } N \text{ rooted;} \\ g^{-1}(x), & \text{if } x\text{'s } f, g\text{-chain is } N' \text{ rooted.} \end{cases} \quad (4)$$

The two rooted cases are illustrated by Figures 6.1 and 6.2. It is straightforward to argue that π is 1-1, onto, and respects f, g-chains. Moreover, since π respects f, g-chains, for each x, we have $x \in A \iff \pi(x) \in B$. The only problem is that π may not be recursive.

The definition of π is based on a global analysis of the structure of f, g-chains. Myhill's construction is much more local. It defines a $\hat{\pi}$ in stages $0, 1, 2, \ldots$. Initially, $\hat{\pi} = \emptyset$.

> *Stage $2x$.* If $\hat{\pi}(x)$ is defined, go on to stage $2x + 1$. Otherwise, traverse the f, g-chain of x forward until an N' vertex y' is encountered that is not yet in range($\hat{\pi}$). Define $\hat{\pi}(x) = y'$.

> *Stage $2x + 1$.* If $\hat{\pi}^{-1}(x')$ is defined, go on to stage $2x + 2$. Otherwise, traverse the f, g-chain of x' forward until an N vertex y is encountered that is not yet in domain($\hat{\pi}$). Define $\hat{\pi}(y) = x'$.

It is clear from the construction that $\hat{\pi}$ is 1-1 and respects f, g-chains. A simple counting argument shows that, for each stage $2x$, a y' as required exists, and, for each stage $2x+1$, a y as required exists. Thus, the even stages of the construction ensure that $\hat{\pi}$ is total, while the odd stages ensure that it is onto. Since f and g are recursive, it is clear from the construction that $\hat{\pi}$ is too. Finally, since $\hat{\pi}$ respects f, g-chains, it is a recursive m-reduction of A to B. Therefore, $\hat{\pi}$ is a recursive isomorphism between A and B as required. □

Proof of Theorem 6.3.2 Let C be a recursively m-complete r.e. set. Let $K_0 = \{\langle i, x \rangle : x \in W_i\}$. Clearly, K_0 is r.e. and, for each i, the function $\lambda x.\langle i, x \rangle$ 1-reduces W_i to K_0. As C is a recursively m-complete r.e. set, there is a recursive m-reduction of K_0 to C. Therefore, it follows that there is a recursive function f such that, for each i, $\varphi_{f(i)}$ recursively m-reduces W_i to C, that is,

$$(\forall i, x)[x \in W_i \iff \varphi_{f(i)}(x) \in C]. \tag{5}$$

Fix an arbitrary i. We use the recursion theorem to construct a special W-program e for the set W_i such that $\varphi_{f(e)}$ recursively *1-reduces* W_i to C. Intuitively, e takes (unfair) advantage of (5) to force $\varphi_{f(e)}$ to be 1-1.

By the recursion theorem for φ, there exists an φ-program e such that, for all x,

$$\varphi_e(x) = \begin{cases} \text{undefined,} & \text{if (a) } (\exists w < x)[\varphi_{f(e)}(w) = \varphi_{f(e)}(x)]; \\ 0, & \text{if not (a) and} \\ & \quad [\text{(b) } \varphi_i(x) \text{ is defined or} \\ & \quad \text{(c) } (\exists y > x)[\varphi_{f(e)}(y) = \varphi_{f(e)}(x)]]; \\ \text{undefined,} & \text{otherwise.} \end{cases} \tag{6}$$

Suppose by way of contradiction that $\varphi_{f(e)}$ is not 1-1. Let x_0 be the *least* number such that, for some $y > x_0$, $\varphi_{f(e)}(y) = \varphi_{f(e)}(x_0)$ and let y_0 be the least such y.

When $x = x_0$, clause (a) fails and clause (c) holds in (6). Hence, $x_0 \in W_e$. (Recall that $W_e = \text{domain}(\varphi_e)$.)

When $x = y_0$, clause (a) holds in (6). Hence, $y_0 \notin W_e$.

However, since $\varphi_{f(e)}(x_0) = \varphi_{f(e)}(y_0)$, by (5) we have $x_0 \in W_e \iff y_0 \in W_e$. But this is a contradiction. Therefore, $\varphi_{f(e)}$ is 1-1.

Now, since $\varphi_{f(e)}$ is 1-1, it follows that, for each x, clauses (a) and (c) in (6) fail to hold. Therefore, $W_e = W_i$. Hence, $\varphi_{f(e)}$ recursively 1-reduces W_i to C.

Therefore, since i was chosen arbitrarily, C is recursively 1-complete for the class of r.e. sets. $\square$

The following corollary formally states what was shown about f in the above proof. The corollary will useful in the proof of Theorem 6.5.1 below.

Corollary 6.3.6 *Suppose C is a recursively m-complete r.e. set and that f is a recursive function such that, for all i, $\varphi_{f(i)}$ recursively m-reduces W_i to C. Then, for each r.e. set A, there is a W-program e for A such that $\varphi_{f(e)}$ recursively 1-reduces A to C.*

LESSONS FROM THE RECURSION THEORETIC RESULTS

In the complexity theoretic work that follows we cannot, in general, make direct use of the above recursion theoretic results and proof techniques. (For a nice exception, see the proof of Theorem 6.5.1.) However, these recursion theoretic results and proofs embody many nice ideas. We briefly consider what sorts of these ideas might be useful for the complexity theory below.

The most obvious things these results have to offer are the ideas behind their proof techniques. These techniques are elegantly simple, and one would expect variants of them to work in complexity theoretic settings. This is the case; ideas from the proofs of this section will play key roles in what follows. But, it is also the case that the complexity theoretic results obtained through these ideas are generally weaker than the results of this section. This is not unexpected. Complexity theory is a much more constrained subject than recursion theory. It is typical in passing from a theory to a more constrained version of it that new distinctions arise, and, in the context of these new distinctions, the analogs of many key results and techniques of the original theory fail or are much more limited in scope. For example, one-way functions have no recursion theoretic analogs and the possible existence of these functions greatly complicate the complexity theoretic situation. For another example, it turns out that the polynomial-time analog of Theorem 6.3.1 is true *if and only if* (as seems unlikely) P = PSPACE—see Theorem 6.4.4 below.

Another use of the results of this section is as a basis for questions. As discussed above, one expects the theory of polynomial-time reductions and complete m-degrees to have many more distinctions than the analogous recursion theory. One way to find these distinctions is to consider complexity theoretic variants of these recursion theoretic results. For example, given a complexity class $\mathcal{C}$, e.g., NP, one might ask the following questions based on Theorem 6.3.2.

Are all the m-complete sets for $\mathcal{C}$ 1-equivalent?
Are all the 1-complete sets for $\mathcal{C}$ 1-li-equivalent?
Are all the 1-li-complete sets for $\mathcal{C}$ 1-li, p-invertible equivalent?
Are all the 1-li, p-invertible-complete sets for $\mathcal{C}$ p-isomorphic?

These sorts of questions have turned out to be productive—at least in our opinion. We have organized this paper around the general principle of considering complexity theoretic versions of the results of this section to see what can be shown true, what can be shown false, and what turns out to be mysterious.

6.4   Sufficient Conditions for P-Isomorphism

6.4.1   COMPLEXITY THEORETIC VERSIONS OF CANTOR-BERNSTEIN

Theorem 6.3.1 provides a sufficient (and necessary) condition for the recursive isomorphism of two sets. In this section we consider sufficient conditions for the polynomial-time isomorphism of two sets. Since the results of this section apply to *arbitrary* sets, the sufficient conditions will turn out to be rather strong. When, instead of arbitrary sets, the sets involved are cylinders or are complete for some complexity class, much weaker sufficient conditions can be obtained as discussed later.

Machtey, Winklmann, and Young [MWY78, Proposition 2.3] did a computational complexity analysis of Rogers's variant of the construction for Theorem 6.3.1.[20] Not surprisingly, their analysis shows that the complexity properties of the unmodified Myhill and Rogers constructions are dreadful. However, Dowd [Dow82] was able to show the following linear-space analog to Theorem 6.3.1 through a construction very much in the same spirit as that in the proof of Theorem 6.3.1.

Theorem 6.4.1 ([Dow82]) *If A and B are recursively 1-equivalent as witnessed by linear length bounded, linear-space computable reductions, then, A and B are linear-space isomorphic.*

In the theory of polynomial-time reducibilities the closest known analog to Theorem 6.3.1 is due to Berman and Hartmanis.

---

[20]Rogers's variant [Rog58] is a relocation of the Theorem 6.3.1 construction into the theory of programming systems.

Theorem 6.4.2 ([BH77]) *If sets A and B are m-equivalent as witnessed by reductions that are (a) 1-1, (b) length increasing, and (c) p-invertible, then A and B are p-isomorphic.*

Proof Suppose N, N', A, B, f, and g are as in the proof of Theorem 6.3.1. Further suppose that f and g satisfy hypotheses (a), (b), and (c). Recall our analysis of the possible structures of f, g-chains from the proof of Theorem 6.3.1. Since f and g are length increasing, we have that each f, g-chain is rooted. So, by the Cantor-Bernstein Theorem construction, π, as defined below, is an isomorphism between A and B.

$$\pi = \lambda x \cdot \begin{cases} f(x), & \text{if } x\text{'s } f, g\text{-chain is } N \text{ rooted;} \\ g^{-1}(x), & \text{if } x\text{'s } f, g\text{-chain is } N' \text{ rooted.} \end{cases}$$

Since f and g are length increasing, for each $z \in (N \cup N')$, there are at most $|z|$ many vertices preceding z in its f, g-chain and all of these vertices are of length less than $|z|$. Thus, since f and g are p-invertible, it follows that, given z, one can find the root of z's f, g-chain in polynomial-time. Hence, π is polynomial-time computable. □

The length increasing and the p-invertibility hypotheses of Theorem 6.4.2 are clearly very strong and it is worth asking whether either of them can be weakened. Watanabe [Wat85] conjectured that, if one-way functions exist, then the p-invertibility hypothesis cannot be weakened. This was confirmed in the following striking result of Ko, Long, and Du.

Theorem 6.4.3 ([KLD87]) $P = UP$ *if and only if every 1-li-degree collapses.*

Proof The $\Longrightarrow$ direction follows from the observation that if one-way functions do not exist, then 1-li-reductions are p-invertible, and, hence, by Theorem 6.4.2, 1-li-equivalent sets are p-isomorphic. The $\Longleftarrow$ direction follows from a pretty and insightful construction, see Theorem 6.6.2 and its proof below. □

One of the reasons that Theorem 6.4.3 is so striking is that it gives a complexity *characterization* of a degree structure property. Theorem 6.4.3 thus essentially settles the question of the general structure of 1-li-degrees. It is a natural question whether there are complexity characterizations of the general collapse of other sorts of degrees, e.g., 1-honest-degrees, 1-degrees, and m-degrees. There has been recent progress on this.

Theorem 6.4.4 ([FKR89]) *The following are equivalent:*
(a) $P = PSPACE$.
(b) Every two 1-equivalent sets are p-isomorphic.
(c) Every two p-invertible equivalent sets[21] are p-isomorphic.

---

[21]That is, sets that are 1-equivalent as witnessed by p-invertible reductions.

Proof Sketch By a proof similar to that for Theorem 6.4.1, one can show that if two sets are recursively 1-equivalent as witnessed by 1-1, polynomial-size bounded, Pspace-computable reductions, then the two sets are Pspace-isomorphic. It follows from this that if P = PSPACE, then every 1-degree collapses. Hence, (a) implies (b). Part (b) trivially implies (c). The proof that (c) implies (a) involves a construction that is partly based on [KLD87], see Theorems 6.6.4 and 6.6.5 below.   □

The equivalence of parts (b) and (c) is surprising and not yet well understood. Theorem 6.4.4 comes close to providing a condition under which the length increasing hypothesis of Berman and Hartmanis's Theorem 6.4.2 is necessary.[22]

For m-degrees one can show, without any assumptions, the existence of noncollapsing m-degrees. See Theorem 6.6.6 below.

Hartmanis established a version of Theorem 6.4.2 in the context of Logspace reductions.

Theorem 6.4.5 ([Har78b]) *If two sets are m-equivalent as witnessed by* Logspace *reductions that are (a) 1-1, (b) length squaring, and (c)* Logspace-*invertible, then the two sets are* Logspace-*isomorphic.*

Using Corollary 6.5.7 below and a modified version of Theorem 6.4.3 one can show that Logspace one-way functions exist *if and only if* the Logspace-invertibility hypothesis of Theorem 6.4.5 is necessary.

6.4.2 CYLINDERS

The two main technical results of the 1977 Berman and Hartmanis paper are Theorem 6.4.2 above and [BH77, Theorem 7], a rough, polynomial-time analog of Theorem 6.3.4. In [MY85] Mahaney and Young improve this latter result to obtain the following exact analog of Theorem 6.3.4.

Theorem 6.4.6 ([BH77] [MY85]) *Two cylinders are m-equivalent if and only if they are p-isomorphic.*

Recall that a *conventional* NP-complete set is, informally, a set that results from a straightforward coding of an NP-complete problem that stems from work in combinatorics, optimization, etc. In their 1977 paper Berman and Hartmanis prove that a number of well known conventional NP-complete sets are paddable, and, moreover, they observed that all of the then known conventional NP-complete sets can also be straightforwardly

---

[22]Note that Theorem 6.4.4 does not rule out the possibility that "length nondecreasing" can replace "length increasing" in the hypothesis of Theorem 6.4.2. We suspect that under a stronger condition than P ≠ PSPACE , the length increasing hypothesis of Theorem 6.4.2 is indeed necessary.

shówn to be paddable. Thus, by Theorem 6.4.6 (or [BH77, Theorem 7]) one obtains Berman and Hartmanis's key observation that all of the known (circa 1977) conventional NP-complete sets are p-isomorphic. To date there still is no satisfactory example of a conventional NP-complete set that cannot be shown paddable.

Proof Sketch of Theorem 6.4.6 The $\Longleftarrow$ direction is obvious. We show the $\Longrightarrow$ direction. Suppose that A and B are m-equivalent cylinders that have associated padding functions $p_A(\cdot,\cdot)$ and $p_B(\cdot,\cdot)$, respectively. We argue that A is 1-li, p-invertible reducible to B. By symmetry, then, it is also that case that B is 1-li, p-invertible reducible to A, and, therefore, by Theorem 6.4.2, that A and B are p-isomorphic.

Since $p_B(\cdot,\cdot)$ is p-invertible, it follows that $p_B(\cdot,\cdot)$ is polynomial honest in the sense that there is a $k > 0$ such that, for all x and y, $|p_B(x,y)| > (|x| + |y|)^{1/k} - k$. Define

$$p'_B = \lambda x, y \,. p_B(x, y \mathbf{01}^{(|x|+|y|+k)^k}),$$

where $y\mathbf{01}^{(|x|+|y|+k)^k}$ denotes the string consisting of y (as a string over $\{\mathbf{0},\mathbf{1}\}$), followed by the symbol $\mathbf{0}$, followed by $(|x| + |y| + k)^k$ many occurrences of the symbol $\mathbf{1}$. It is straightforward to argue that $p'_B(\cdot,\cdot)$ is a padding function for B, and that p'_B is length increasing in the sense that, for all x and y, $|p'_B(x,y)| > |x| + |y|$. Define $f' = \lambda x \,. p'_B(f(x),x)$. Since f is an m-reduction of A to B and since p'_B is a padding function for B, it follows that f' is an m-reduction of A to B. Since p'_B is 1-1, length increasing, and p-invertible, it follows from our definition of f' that it has these properties too. $\square$

Cylinders are of independent interest beyond Berman and Hartmanis's observation that the "natural" NP-complete sets are paddable. The following proposition shows that the m-complete sets that have essentially the strongest reducibility properties turn out to be cylinders.

Proposition 6.4.7 *Suppose* C *is a complexity class that: (i) contains* N, *(ii) is closed downward under m-reductions, and (iii) has a complete m-degree. Then, (a) and (b) hold.*
    (a) There is a cylinder that is 1-li, p-invertible complete for C.
    (b) All of the 1-li, p-invertible complete sets for C *are cylinders.*

Before we prove Proposition 6.4.7 we note (without proof) an easy corollary of Theorem 6.4.6.

Corollary 6.4.8 *For each cylinder* B, *a set* A *is m-reducible to* B *if and only if* A *is 1-li, p-invertible reducible to* B.

Proof Sketch of Proposition 6.4.7 Suppose C is an m-complete set for $\mathcal{C}$. Let $C' = \{\langle c, z \rangle : c \in C \ \& \ z \in N\}$. It is straightforward to show that C' is a cylinder and an m-complete set for $\mathcal{C}$. By Corollary 6.4.8 we have that C' is a 1-li, p-invertible complete set for $\mathcal{C}$. Therefore, part (a) follows. Part (b) follows from part (a), Corollary 6.4.8, Theorem 6.4.2, and the easily verified fact that p-isomorphisms map cylinders to cylinders. □

Every m-degree contains a cylinder, but there are 1-degrees that do not (see Theorem 6.6.6 below.) We say a 1-degree is *cylindrical* if and only if the 1-degree contains a cylinder. By Proposition 6.4.7(a) all complete 1-degrees of complexity classes are cylindrical. Now, Theorem 6.4.4 gives P = PSPACE as a sufficient condition for every 1-degree to collapse. The next proposition gives an ostensibly weaker sufficient condition for the collapse of every cylindrical 1-degree.

Proposition 6.4.9 *If A is a cylinder and B is a set 1-equivalent to A, then A and B are* Ptime^{NP}*-isomorphic.*

Thus, P = NP *implies that cylindrical 1-degrees collapse.*

The proof of this proposition is an implicit part of the proof of [MWY78, Theorem 2.6]. We do not know whether the collapse of every cylindrical 1-degree implies P = NP. We know of presumably weaker hypotheses than P = NP that imply that cylindrical 1-degrees collapse to 1-li-degrees, but it is open whether there is a complexity characterization of this sort of collapse.

We note in passing that the obvious Logspace analogs of Theorem 6.4.6 and Proposition 6.4.9 hold.

6.5 Complete Degrees

We now directly address the problem of the structure of complete degrees. Work on this topic has produced some strong positive results. Among these are that

- the complete m-degrees of both RE and NEXP collapse to 1-degrees (Theorems 6.5.1 and 6.5.3 below);

- the complete m-degree of EXP collapses to a 1-li-degree (Theorem 6.5.2 below);

- there is a structural characterization of the collapse/noncollapse of the complete m-degree of EXP (Theorem 6.5.11 below); and

- the existence of sufficiently strong one-way functions (i.e., scrambling functions) implies the noncollapse of the complete m-degrees of NP, PSPACE, EXP, NEXP, and RE (Theorem 6.5.10 below).

However, there are also results which carry the cautionary message that any resolution of the collapse/noncollapse question of the complete m-degree of NP, or PSPACE, or EXP, or ... will also resolve some major complexity class problem. For example consider the complete m-degree of RE. The noncollapse of this degree implies $P \neq NP$ (Proposition 6.4.9) and the collapse of this degree implies the nonexistence of scrambling functions (Theorem 6.5.10). In the face of these "negative" results, it is not a realistic research goal to try to push toward an out right resolution of these complete set problems. A more reasonable goal is to instead try to work toward establishing complexity characterizations of the collapse/noncollapse of particular complete degrees. Such characterizations are potentially very informative. Theorem 6.5.11 below (due to Ganesan and Homer) is a first step in this direction.

6.5.1   PROOFS OF PARTIAL COLLAPSE

Below we sketch proofs that the complete m-degrees of RE and NEXP collapse to 1-degrees and that the complete m-degree of EXP collapses to a 1-li-degree. These are pithy proofs which seem to get at the heart of the matters. Following these proofs we also explain why the nice ideas behind these proofs seem to fail in the context of NP and PSPACE.

Our development of Theorems 6.5.1, 6.5.2, and 6.5.3 below parallels Ganesan and Homer's [GH89] (in which Theorem 6.5.3 is introduced) and we refer the reader to that paper for a good alternative treatment of these theorems. Our proofs and Ganesan and Homer's share many key elements, but differ in points of view.

We begin by considering the complete m-degree of RE.

Theorem 6.5.1 ([Dow78]) *The complete m-degree of RE consists of a 1-degree.*

This theorem first appeared in Dowd's [Dow78]. A cleaner proof of this theorem appears in [GH89]. Our proof is simple, direct application of Corollary 6.3.6. The Dowd proof, the Ganesan and Homer proof, and the proof below all use essentially the same diagonalization trick.

Proof Let C be a polynomial-time m-complete r.e. set. By a straightforward argument, there is a recursive function f such that, for each i, $\varphi_{f(i)}$ is polynomial-time computable and $\varphi_{f(i)}$ m-reduces W_i to C. Fix an arbitrary r.e. set A. By Corollary 6.3.6 there is a W-program e for A such that $\varphi_{f(e)}$ recursively 1-reduces A to C. But, by assumption, $\varphi_{f(e)}$ is polynomial-time computable. Hence, A is (polynomial-time) 1-reducible to C. Since A was chosen arbitrarily, C is (polynomial-time) 1-complete.   □

We next consider the complete m-degree of EXP.

Theorem 6.5.2 ([Ber77]) *The complete m-degree of* EXP *consists of a 1-li-degree.*

This result first appeared in Berman's 1977 Ph.D. dissertation. Watanabe published a very clean, clear proof of this result in [Wat85]. (Also see [GH89, Theorem 1].) In [Wat85] Watanabe also shows that the complete m-degree of each of the classes

- DTIME($2^{\mathcal{O}(n^{1/k})}$), where $k > 0$, and

- DSPACE($s(n)$), where s is space constructible and super-polynomial,

consists of a 1-li-degree.

All of the known proofs of Theorem 6.5.2 use essentially the same diagonalization techniques. In our proof below, to diagonalize against m-reductions that are not 1-1, we use a version of the diagonalization trick employed in the proofs of Theorems 6.3.2 and 6.5.1 above. (See clause (a) in (7) below.) To diagonalize against reductions that are not length increasing we do the following. Suppose that f is a potential m-reduction of E_{e_0}, the set we are constructing, to C, a predetermined complete set. Also suppose that $|f(x_0)| \leq |x_0|$ for some x_0. Then, to diagonalize against f at x_0, we put x_0 into E_{e_0} if and only if $f(x_0)$ is *not* in C. (See clause (b) in (7) below.) There are two key points here. First, the diagonalization depends on the fact that one can *decide* the question "$y \in C$?" in time exponential in $|y|$. Thus, this sort of diagonalization does not work for classes like RE and NEXP which are either not closed or not known to be closed under complements. The second key point is that since $|f(x_0)| \leq |x_0|$, one can decide the question "$f(x_0) \in C$?" within a uniform $2^{\text{poly}(|x_0|)}$ time bound *independent of the reduction* f. Since the construction has to be able to diagonalize against all possible m-reductions f that fail to be length increasing, this point is critical to making the construction work in exponential time.

Proof Sketch In the following, for each $A \in$ EXP, $A(\cdot)$ will denote the (exponential time computable) characteristic function of A. Let C be an m-complete set for EXP. For all i, let $E_i = \psi_i^{-1}(C)$. By our discussion in S 6.2, $\langle E_i \rangle_{i \in N}$ is a programming system for EXP for which the recursion theorem holds. Fix an arbitrary set $A \in$ EXP. Define, for all e and x,

$$D(\langle e, x \rangle) = \begin{cases} 1 - D(\langle e, w_0 \rangle), & \text{if (a) for some } w < x, \ \psi_e(w) = \\ & \quad \psi_e(x) \text{ and } w_0 \text{ is the least} \\ & \quad \text{such } w; \\ 1 - C(\psi_e(x)), & \text{if (b) } |\psi_e(x)| \leq |x| \text{ and not (a);} \\ A(x), & \text{otherwise.} \end{cases}$$

A straightforward argument shows that D is in EXP. (Recall that the function $\lambda i, x . \psi_i(x)$ is computable in exponential time.) Hence, by the recursion theorem for $\langle E_i \rangle_{i \in N}$ there is an E-program e_0 such that $E_{e_0}(\cdot) = \lambda x . D(\langle e_0, x \rangle)$. Thus, for all x,

$$
E_{e_0}(x) = \begin{cases} 1 - E_{e_0}(w_0), & \text{if (a) for some } w < x, \ \psi_{e_0}(w) = \\ & \psi_{e_0}(x) \text{ and } w_0 \text{ is the least} \\ & \text{such } w; \\ 1 - C(\psi_{e_0}(x)), & \text{if (b) } |\psi_{e_0}(x)| \le |x| \text{ and not (a);} \\ A(x), & \text{otherwise.} \end{cases} \tag{7}
$$

Suppose by way of contradiction that ψ_{e_0} is not 1-1. Let x_0 be the least number such that, for some $w < x_0$, $\psi_{e_0}(w) = \psi_{e_0}(x_0)$. Then, by the case of clause (a) in (7), $w \in E_{e_0} \iff x_0 \notin E_{e_0}$. But, since ψ_{e_0} is an m-reduction of E_{e_0} to C and since $\psi_{e_0}(w) = \psi_{e_0}(x_0)$, we have that $w \in E_{e_0} \iff x_0 \in E_{e_0}$, a contradiction. Therefore, ψ_{e_0} is 1-1.

Suppose by way of contradiction that ψ_{e_0} is not length increasing. Let x_0 be the least number such that $|\psi_{e_0}(x_0)| \le |x_0|$. Then, by the case of clause (b) in (7), $x_0 \in E_{e_0} \iff \psi_{e_0}(x_0) \notin C$. But, since ψ_{e_0} is an m-reduction of E_{e_0} to C, this is a contradiction. Therefore, ψ_{e_0} is length increasing.

By our definition of the E_i's, ψ_{e_0} is an m-reduction of E_{e_0} to C. Hence, E_{e_0} is 1-li-reducible to C.

Since ψ_{e_0} is 1-1 and length increasing, for every x, clauses (a) and (b) fail to hold in (7). Hence, $E_{e_0} = A$, and, therefore, A is 1-li-reducible to C.

Since A was an arbitrary member of EXP, we thus have that C is 1-li-complete for EXP. $\qquad \square$

By a clever combination of the ideas behind the proofs of Theorems 6.5.1 and 6.5.2, Ganesan and Homer established

Theorem 6.5.3 ([GH89]) *The complete m-degree of* NEXP *consists of a 1-degree. In fact, every two m-complete sets for* NEXP *are 1-equivalent as witnessed by exponentially honest reductions.*

They also show the analogous result for the classes NTIME($2^{\mathcal{O}(n^{1/k})}$), where $k > 0$.

Proof Sketch of Theorem 6.5.3 To keep to the argument simple, we shall not worry about exponential honesty and show only that the m-complete sets of NEXP are 1-complete.

In the following, for each set A, $A(\cdot)$ will denote the partial function that is 1 on each element of A and undefined otherwise.

Let C be an NEXP-complete set. For all i, let $E_i = \psi_i^{-1}(C)$. By our discussion in S 6.2, $\langle E_i \rangle_{i \in N}$ is a programming system for NEXP for which the recursion theorem holds.

Fix an $A \in$ NEXP. Since NEXP is contained in deterministic double exponential time, it is straightforward to argue that there is a polynomial

p such that, for all x and all w such that $2^{p(|w|)} < |x|$, one can deterministically evaluate $\overline{A}(w)$ within a uniform $\mathcal{O}(2^{|x|})$ time bound.

By the recursion theorem for $\langle E_i \rangle_{i \in N}$ there is an E-program e_0 such that, for all x,

$$
E_{e_0}(x) = \begin{cases}
\overline{A}(w), & \text{if (a) } (\exists w < x)[\, 2^{p(|w|)} < |x| \ \& \\
& \qquad \psi_{e_0}(w) = \psi_{e_0}(x)\,]; \\[4pt]
\text{undefined}, & \text{if not (a) and} \\
& \quad \text{(b) } (\exists w < x)[\, |x| \le 2^{p(|w|)} \ \& \\
& \qquad \psi_{e_0}(w) = \psi_{e_0}(x)\,]; \\[4pt]
1, & \text{if not (a) and not (b) and} \\
& \quad [\text{(c) } x \in A \text{ or} \\
& \quad \text{(d) } (\exists y > x)[\, |y| \le 2^{p(|x|)} \ \& \\
& \qquad \psi_{e_0}(x) = \psi_{e_0}(y)\,]\,]; \\[4pt]
\text{undefined}, & \text{(e) otherwise.}
\end{cases}
\tag{8}
$$

We note that the right hand side of (8), as a function of $\langle e_0, x \rangle$, corresponds to an NEXP set since:

- clause (a) is an EXP test and, by our assumption on p, one can deterministically compute $\overline{A}(w)$ within a uniform $\mathcal{O}(2^{|x|})$ time bound;

- clause (b) is also an EXP test; and

- clauses (c) and (d) are NEXP tests.

So, we can indeed apply the recursion theorem for $\langle E_i \rangle_{i \in N}$ to the set corresponding to this function of $\langle e_0, x \rangle$.

The case of clause (a) in (8) guarantees that there are no x and x' with $x' < x$ such that $2^{p(|x'|)} < |x|$ and $\psi_{e_0}(x') = \psi_{e_0}(x)$. The cases of clauses (b) and (d) in (8) guarantee that there are no x and x' with $x' < x$ such that $|x| \le 2^{p(|x'|)}$ and $\psi_{e_0}(x) = \psi_{e_0}(x')$. (We leave these two arguments to the reader.) Therefore, we have that ψ_{e_0} is a 1-reduction of E_{e_0} to C. Now, since ψ_{e_0} is 1-1, clauses (a), (b), and (d) fail to hold for each and every x. Hence, it follows by the cases of clauses (c) and (e) that $E_{e_0} = A$. Therefore, ψ_{e_0} is a 1-reduction of A to C. Since A was chosen arbitrarily, the theorem follows. $\qquad \square$

In contrast to the situation for RE, NEXP, and EXP, there are essentially no details known about the structure of the complete m-degrees of either NP or PSPACE. The proof techniques used for Theorems 6.5.1, 6.5.2, and 6.5.3 do not seem applicable in the context of either NP or PSPACE and there are no presently known alternatives to these techniques. The reason for the apparent failure of these techniques is this. The constructions for Theorems 6.5.2 and 6.5.3 each makes strong use of the fact that $\lambda i, x . \psi_i(x)$

is computable in time exponential in $|i|+|x|$. There is no known polynomial-space computable universal function for Ptime and the existence of such a function seems doubtful. (Kozen [Koz80] presents some evidence against the existence of such functions.) However, for the sake of comparison with the previous theorems of this section we note

Proposition 6.5.4 *Suppose ψ' is a universal function for* Ptime *such that the two sets*

$$\{\langle i,j,x \rangle : \psi'_i(x) = \psi'_j(x)\} \qquad \{\langle i,w,x \rangle : |w| \leq |x| \ \& \ w = \psi'_i(x)\}$$

are both in PSPACE. *Then, the complete m-degree of* PSPACE *consists of a 1-li-degree.*

An analogous result holds for NP when one makes the additional assumption that NP = coNP.

The proof of Proposition 6.5.4 is a modification of that for Theorem 6.5.2—with a proviso: ψ' may be *so* unnatural that the recursion theorem may fail for it. To obtain the proposition in this case, one is reduced to modifying one of the proofs of Theorem 6.5.2 that avoids use of the recursion theorem, see [Wat85] and [GH89] for examples.

Although there may be no polynomial space computable universal functions for Ptime, there is indeed a universal function for Logspace, θ, such that that the two sets

$$\{\langle i,j,x \rangle : \theta_i(x) = \theta_j(x)\} \qquad \{\langle i,w,x \rangle : |w| \leq |x| \ \& \ w = \theta_i(x)\}$$

are both in PSPACE. The proof Proposition 6.5.4 can thus be modified to obtain the following nice analog of Theorem 6.5.2 for PSPACE.

Theorem 6.5.5 ([Rus86]) *Relative to* Logspace *reductions, the complete m-degree of* PSPACE *consists of a 1-li-degree.*

The above theorem can be improved a bit by observing that for θ as above, given any fixed polynomial p, the set

$$\{\langle i,w,x \rangle : |w| \leq p(|x|) \ \& \ w = \theta_i(x)\}$$

is also in PSPACE. So, by a slight modification of the proof of Proposition 6.5.4, one obtains

Theorem 6.5.6 *Relative to* Logspace *reductions, the complete m-degree of* PSPACE *consists of a 1-length-squaring-degree.*

Theorems 6.4.5 and 6.5.6 together yield

Corollary 6.5.7 *Relative to* Logspace *reductions, if one-way functions do not exist, then the complete m-degree of* PSPACE *collapses.*

Allender showed the following related result by techniques similar to the ones discussed above.

Theorem 6.5.8 ([All88]) *All the members of the 1-L complete*[23] *degree for* PSPACE *are p-isomorphic.*

6.5.2   CONSEQUENCES OF COLLAPSE/NONCOLLAPSE

None of the results of the previous subsection completely settles the question of the structure of any complete m-degree. For instance, in the case of RE, it is known that its complete m-degree collapses to a 1-degree, but the extent to which this degree collapses further (if at all) is open. These questions are likely to be hard—an answer to any of them would have some profound complexity theoretic implications. The next proposition summarizes some of these.

Proposition 6.5.9 *(a) The noncollapse of the complete m-degree of* EXP *implies* $P \neq UP$.
    (b) The noncollapse of the complete m-degree of RE *implies* $P \neq NP$.
    (c) The noncollapse of the complete m-degree of NEXP *implies* $P \neq NP$.
    (d) The noncollapse of the complete m-degree of PSPACE *implies* LOG-SPACE $\neq$ NP.
    (e) The collapse of the complete m-degree of NP *implies* $P \neq NP$.

Proof Sketch Part (a) follows by Theorems 6.4.3 and Theorem 6.5.2. Using Propositions 6.4.9 and 6.4.7(a), part (b) follows from Theorem 6.5.1 and part (c) by Theorem 6.5.3. Part (d) follows by Proposition 6.4.9 and Theorem 6.5.5. Part (e) follows from the observation that if $P = NP$, then the set $\{1\}$ is NP-complete, but clearly not p-isomorphic to SAT. $\square$

Until recently, Proposition 6.5.9(e) was the only known result that provided a "complexity theoretic" consequence of the *collapse* of a m-complete degree. The next theorem (obtained in 1988) implies that if any of the standard complete m-degrees collapse, then there is a limit on the power of one-way functions. Recall that a scrambling function is a one-way function such that, for all nonempty A, $f(A)$ is nonpaddable.

Theorem 6.5.10 ([KMR89]) *If scrambling functions exist, then the complete 1-li-degrees of each of* NP, PSPACE, EXP, NEXP, *and* RE *all fail to collapse.*
    Thus, if any of the 1-li-complete degrees of NP, PSPACE, EXP, NEXP, *and* RE *collapse, then scrambling functions do not exist.*

---

[23]A *1-L reduction* [HIM81] is roughly an m-reduction that is computable by a logspace bounded Turing machine that has a one-way input head.

Disproving the existence of scrambling functions. is likely to be hard because these functions exist relative to random oracles, see Theorem 6.7.5 below. Thus, proving that any of the complete m-degrees of NP, PSPACE, EXP, NEXP, and RE collapse is also probably very hard.

Proof Sketch of Theorem 6.5.10   *Terminology.* A complexity class C is *image complete* if and only if, for each f, a 1-1, length increasing, polynomial-time computable function, and each A, an m-complete set for C, $f(A)$ is also an m-complete set for C. It is straightforward to argue that each of NP, PSPACE, EXP, NEXP, and RE is image complete.

Ko, Long, and Du [KLD87] show that if one-way functions exist, then so do length increasing one-way functions. The same argument applies to scrambling functions. So, suppose f is a scrambling function which we assume is length increasing. Suppose also that C is an image complete complexity class and that C_0 is a paddable, 1-li-complete set for C. (By Proposition 6.4.7(a) such a C_0 must exist.) Consider $f(C_0)$. Since C is image-complete, $f(C_0)$ is m-complete for C. Since C_0 is l-li-complete for C and since f is 1-1 and length increasing, we have that $f(C_0)$ is 1-li-complete for C. As f is a scrambling function, $f(C_0)$ is not paddable. Therefore, since p-isomorphisms preserve paddability, it follows that C_0 and $f(C_0)$ are two 1-li-complete sets that are not p-isomorphic.   □

It is natural to ask whether there are complexity theoretic characterizations of the collapse/noncollapse of any complete m-degrees. Watanabe [Wat85] conjectured the converse of Proposition 6.5.9(a). Machtey, Winklmann, and Young [MWY78, p. 51] in effect conjectured the converse of Proposition 6.5.9(b). (Their actual conjecture was an analogous statement for acceptable programming systems for the class of partial recursive functions.) Theorem 6.5.10 is a very partial confirmation of these two conjectures. The only characterization of the collapse/noncollapse of a complete m-degree is the following result due to Ganesan and Homer.

Theorem 6.5.11 ([Gan89]) *The complete m-degree of* EXP *is noncollapsing if and only if there exist* C, *an m-complete set for* EXP, *and a one-way function* f *for which there is no p-invertible function* g *such that* $g(C) \subseteq f(C)$ *and* $g(\overline{C}) \subseteq f(\overline{C})$.

Theorem 6.5.11 is not a "complexity theoretic" characterization of the collapse of EXP's complete m-degree in the sense that the converse of Proposition 6.5.9(a) would be. The theorem is more of an analysis which shows, if this complete degree fails to collapse, how the failure must occur. (Watanabe [Wat88] has results related to this.)

Proof Sketch of Theorem 6.5.11   If for each C, an m-complete set for EXP, and each 1-1, length increasing f there *is* a p-invertible g such that $g(C) \subseteq f(C)$ and $g(\overline{C}) \subseteq f(\overline{C})$, then, by a straightforward application of Theorems 6.4.2 and 6.5.2, we have that the complete degree of EXP collapses. Thus, the $\Longrightarrow$ direction follows.

To show the $\Longleftarrow$ direction, first suppose that C and f are as in the hypothesis. If C is not paddable, we are done. So, suppose that C is paddable. By a diagonalization argument one can construct A, an m-complete set for EXP such that (i) $f(C) \subseteq A$, (ii) $f(\overline{C}) \subseteq \overline{A}$, and (iii) for any f' that 1-li-reduces C to A, it is the case that both $f'(C) - A$ and $f'(\overline{C}) - \overline{A}$ are finite. (We leave the details of the construction to the reader.) Suppose by way of contradiction that there is a p-invertible reduction from C to A. Using C's paddability, it follows that there must exist g, a length increasing, p-invertible reduction of C to A. Hence, $g(C) - f(C)$ and $g(\overline{C}) - f(\overline{C})$ are finite. Using C's paddability again, it follows that there is a length increasing, p-invertible g' such that $g'(C) \subseteq f(C)$ and $g'(\overline{C}) \subseteq f(\overline{C})$, a contradiction. $\qquad\square$

6.6   Degree Structure

One of many things left open by the work of the previous section is the general question of which of the many potential structures of degrees actually occur. For example, are there *any* m-degrees that collapse—much less complete ones? There are two main motivations for construction of concrete examples of degrees with a particular structural property: (i) to see what sort of hypotheses (if any) are required for such degrees to exist, and (ii) to possibly obtain hints on how to show that some complete degree has the property in question. Taking the example of collapsing degrees again, at first blush it is conceivable that the existence of collapsing degrees might require a hypothesis like P = NP or P = UP. If this were the case, then the collapse of any m-degree would be highly suspect. Well, this is not the case as shown by

Theorem 6.6.1 ([KMR88]) *Collapsing degrees exist. Moreover, there is a collapsing degree that is 2-tt-complete*[24] *in EXP.*

The proof of this theorem is a finite injury priority argument which is too involved to sketch here, see [KMR88] for the details. One of our (thoroughly naïve) motivations for working on collapsing degrees was to make progress

---

[24] A set A is *2-truth-table (2-tt) reducible to* B if and only if there exists polynomial-time computable f such that (i) for each x, $f(x)$ codes both a binary boolean function α and a pair of numbers (x_1, x_2), and (ii) for all x, $x \in A$ if and only if $\alpha([x_1 \in B?], [x_2 \in B?])$. **N.B.** The relation of 2-tt-reducibility is *not* transitive. Hence, it is an abuse of terminology to call "2-tt-reducibility" a reducibility. Furthermore, while the notion of 2-tt-complete makes perfect sense, the notion of a 2-tt-degree does not!

toward a proof that the complete m-degree of EXP does collapse. However, in our constructions of collapsing degrees, 2-tt-complete for EXP is as close to m-complete for EXP as we have been able to manage. Another of our motivations for showing Theorem 6.6.1 was to provide a counterpart for the following beautiful result of Ko, Long, and Du.

Theorem 6.6.2 ([KLD87]) *Suppose one-way functions exist. Then, there is a noncollapsing 1-li-degree, i.e., there are 1-li-equivalent sets that are not p-isomorphic. Moreover, there are such sets that are also 2-tt-complete for EXP.*

As discussed in S 6.4.1, the existence of one-way functions is a necessary condition for the existence of noncollapsing 1-li-degrees, and, Theorem 6.6.2 shows the condition is also sufficient. (See the discussion around Theorem 6.4.3 above.) One of Ko, Long, and Du's goals was to try to confirm Watanabe's conjecture that the complete m-degree of EXP collapses if and only if one-way functions exist. However, as with our Theorem 6.6.1, 2-tt-complete was as close they could get to m-complete. Pushing closer to m-complete than 2-tt-complete for either of Theorems 6.6.1 or 6.6.2 seems to be a very hard and will probably require radical, new techniques.

The proof of Theorem 6.6.2 builds on Berman and Hartmanis's proof of Theorem 6.4.2 and introduces some important new ideas.

Proof Sketch of Theorem 6.6.2 Suppose one-way functions exist. Then, by [KLD87] there exists t, a *length increasing* one-way function. We define $f: N \to N'$ by the following two equations.

$$f(2x) = 4t(x) + 1. \qquad f(2x + 1) = 4x + 3.$$

Let g have the same definition as f except that we regard g as a function from N' to N. Clearly, f and g are 1-1 and length increasing.

Terminology. A function $h: N \to N'$ *crosses* an f, g-chain C if and only if for some x, an N vertex of C, $h(x)$ is not an N' vertex of C.

Lemma 6.6.3 *Suppose $h: N \to N'$ is p-invertible. Then, h crosses infinitely many f, g-chains.*

Proof of the Lemma Let p be a nondecreasing polynomial such that, for all x, $|h(x)| \le p(|x|)$. For each y, let A_y be the set of N' vertices of the f, g-chain of $(4y + 1)'$ that are of length $\le p(|4y + 1|)$. Since f and g are 1-1, length increasing, polynomial-time computable functions, there are fewer than $p(|4y + 1|)$ many elements of A_y, and, in fact, there is an obvious polynomial-time procedure that, given y, lists all the elements of A_y. By our definitions of f and g, we have that, for all x,

$$h(2x) \text{ is in the same chain as } 2x \iff h(2x) \text{ is in } A_{t(x)}. \qquad (9)$$

See Figure 6.3.

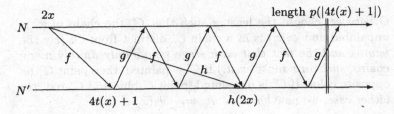

FIGURE 6.3. $h(2x)$ lands in $A_{t(x)}$.

Now, suppose by way of contradiction that the lemma is false. Then, it follows from our definitions of f and g that, for all but finitely many x, $2x$ and $h(2x)$ are in the same chain. Let

$$y_0 = n-1(\{t(x) : 2x \text{ and } h(2x) \text{ are in distinct chains}\}).$$

Then, by (9) and our choice of y_0,

$$(\forall y > y_0)[t^{-1}(y) \text{ is defined} \iff (\exists z' \in A_y)[h(2t^{-1}(y)) = z']]. \quad (10)$$

Observe that:

$$h(2t^{-1}(y)) = z'$$

$$\iff t^{-1}(y) = \frac{1}{2}h^{-1}(z') \quad (11)$$

$$\iff t(\frac{1}{2}h^{-1}(z')) = y. \quad (12)$$

Since t and h^{-1} are polynomial-time computable, given y and z', one can in polynomial (in $|y| + |z'|$) time check whether $h(2t^{-1}(y)) = z'$ using (12), and, if $h(2t^{-1}(y)) = z'$, compute $t^{-1}(y)$ using (11). Since one can list all the elements of A_y in polynomial (in $|y|$) time, it follows by (10) that, for each $y > y_0$, one can determine $t^{-1}(y)$ in polynomial in $|y|$ time. Therefore, t is p-invertible, a contradiction. $\quad\Box$ **Lemma**

Using the lemma, we noneffectively construct 1-li-equivalent sets $A \subseteq N$ and $B \subseteq N'$ that are such that no p-invertible function m-reduces A to B. The construction is in stages $0, 1, \ldots$. In each stage, the construction "paints" a finite number of f, g-chains blue or red. A chain painted blue has its N vertices in A and its N' vertices in B, and a chain painted red has its N vertices in $\overline{A}$ and its N' vertices in $\overline{B}$. Since A and B will be constructed to respect f, g-chains, we shall have that f 1-li-reduces A to B and g 1-li-reduces B to A. The construction eventually paints all chains and never repaints a chain. Initially, all chains are unpainted.

Stage i. First, paint red each unpainted chain that has a vertex less than i. Next, if ψ_i is not p-invertible, go on to stage $i + 1$.

> Otherwise, choose the least x such that C, the chain of x, is
> unpainted and $\psi_i(x)$ is in a chain C' distinct from C. (By the
> lemma and the fact that each stage paints only finitely many
> chains, such an x must exist.) If C' is painted, then paint C the
> opposite color. If C' is unpainted, paint C blue and C' red. (In
> either case, we now have $x \in A \iff \psi_i(x) \notin B$.)

Clearly, A and B are as required.

It is fairly straightforward to make the construction of A and B effective
and with careful programming the construction can be made to produce A
and B in EXP. Making A and B in addition 2-tt-complete for EXP is a
simple, clever trick which we refer the reader to [KLD87] for the details.

$\square$

By Theorem 6.5.2, if one-way functions exist, there are two possible
structures of the complete m-degree for EXP: either the degree collapses or
it contains sets that are 1-li-equivalent but not p-isomorphic. By Theorems
6.6.1 and 6.6.2, if one-way functions exist, both possible degree structures
are realized within the sets 2-tt-complete for EXP.[25]

By Theorems 6.5.1 and 6.5.3 the m-degrees of both RE and NEXP collapse to 1-degrees. However, at present it seems possible that these degrees
may contain sets that are not even m-honest-equivalent. Fenner has shown
that, if P $\neq$ PSPACE, there are 1-degrees that have this structure.

Theorem 6.6.4 ([Fen89] [FKR89]) *Suppose* P $\neq$ PSPACE. *Then there
is a noncollapsing 1-degree. In fact, there exist 1-equivalent sets that fail
to be m-honest-equivalent. Moreover, there are such sets that are also 2-tt-
complete for* EXP.

The proof of this theorem is based in part on ideas from the Ko, Long,
and Du construction and on Bennett's work on reversible Turing Machines
[Ben89].

In related work, Fenner, Kurtz, and Royer show that, if P $\neq$ PSPACE,
then there are 1-degrees that have a very strange failure of collapse.

Theorem 6.6.5 ([FKR89]) *Suppose* P $\neq$ PSPACE. *Then there is a noncollapsing p-invertible degree, i.e., there exist p-invertible-equivalent sets
that fail to be p-isomorphic. Moreover, there are such sets that are also
2-tt-complete for* EXP.

---

[25]Theorem 6.6.2 shows the existence of a noncollapsing 1-li-degree provided
one-way functions exist. We suspect that by combining the techniques of the
proofs of Theorems 6.6.1 and 6.6.2 one can show that, if one-way functions exist, then there is an m-degree that collapses to a 1-li-degree, but not to a p-
isomorphism type.

At present essentially nothing is known about the complete m-degrees of NP and PSPACE. It is possible that these degrees contain sets that are not 1-equivalent. M-degrees that contain sets that fail to be 1-equivalent do exist in "small" complexity classes as shown by

Theorem 6.6.6 ([KMR87a]) *(a) Suppose t is fully time constructible (see [HU79] for the definition) and* $P \subset DTIME(t(n))$. *Then, there exists a noncollapsing m-degree in* $DTIME(t(n)) - P$.

(b) There exists a 2-tt-complete m-degree in EXP that contains infinitely many distinct 1-degrees.

Proof Sketch *Terminology.* A set A is *p-subset-immune* if and only if A is infinite and it contains no infinite, polynomial-time decidable subsets. A is *p-enumeration-immune* if and only if A is infinite and for each 1-1, polynomial-time computable g, we have range$(g) \not\subseteq A$.

It is easy to show that if an m-degree contains a p-subset-immune set, then the m-degree is noncollapsing, and if the m-degree contains a p-enumeration-immune set, then the degree is made up of infinitely many different 1-degrees. It follows by a result of Geske, Huynh, and Seiferas [GHS89, Theorem 4] that, for each t as in the hypothesis of part (a), there is a p-subset-immune set in $DTIME(t(n)) - P$. Hence, part (a) follows. By a fairly straightforward diagonalization one can construct a p-enumeration-immune set that is 2-tt-complete for EXP. Hence, part (b) follows also. $\square$

The above results have been concerned with constructing degrees that exhibit a certain amount of collapsing. For each of the noncollapsing cases we have been content to simply build two sets that witness the form of non-collapse of interest, e.g., two non-1-li-equivalent sets in the case of Theorem 6.6.2. One can also consider the problem of describing the internal structure of noncollapsing degrees. Mahaney [Mah81] showed that every noncollapsing m-degree contains an $\omega + 1$ chain of sets ordered under 1-li-reductions none of which sets are p-isomorphic to any of the others. Mahaney and Young [MY85] later extended this result to

Theorem 6.6.7 ([MY85]) *In each noncollapsing m-degree, any countable partial ordering can be realized as a collection of pairwise non-p-isomorphic sets ordered under 1-li-reductions.*

6.7 Relativization Results

Thus far there has been little substantive said about degrees, complete or otherwise, in NP and PSPACE. This is because there is not much to report. Results about such degrees seem very hard to come by and seem to be beyond conventional techniques. Relativizations can be used to give a partial explanation of why this is the case.

The purpose of relativization results is to informally establish that a given property is independent of the standard diagonalization and simulation techniques of recursion and complexity theories, see [BGS75]. The idea is this. The aforementioned techniques seem to be indifferent to the presence of oracles in models of computation. That is, if with these techniques one can prove some property P of a standard model of computation, say Turing machines, then it seems to be the case that, one can also prove that P holds for Turing machines with an arbitrary oracle. Thus, if there is an oracle A for which *not P* holds relative to A, it seems unlikely that these standard techniques can provide a proof of P in the unrelativized setting. Baker, Gill and Solovay [BGS75] illustrate this with oracles A and B such that (i) relative to A we have P = NP and (ii) relative to B we have P $\neq$ NP. The interpretation of these two results is that the resolution of the P versus NP question is beyond the scope of the standard diagonalization and simulation techniques. However, since we have not precisely defined what these standard techniques are, relativization results are only an *informal* method of independence, and, thus, we cannot formally prove that all such methods will relativize. On the other hand, most proofs using these methods do indeed relativize.

Another use of relativization results is to establish the plausibility of certain hypotheses. One can interpret computability relative to a given oracle as a computational "possible world." Arguing the plausibility of some fact based on an oracle result is usually a pretty tenuous proposition. However, when the oracle in a relativatizion result is drawn from certain special subclasses, arguing plausibility from an oracle result has somewhat better support. We discuss this issue for the classes of sparse and random oracles below.

For relativization results about the structure of m-degrees one needs be careful about which machines (i.e., those for language acceptors, m-reductions, and isomorphisms) have access to the oracle. Relativized settings in which both language acceptors and machines computing reductions have access to the oracle are called *full* relativizations and those settings in which just the language acceptors have access to the oracle are called *partial* relativizations [Rog67, S 9.3]. Partial relativizations are of interest when studying the properties of *actual* m-reductions and p-isomorphisms. For "independence" results as discussed above, full relativizations seem to be the appropriate setting as machines accepting languages and computing reductions are treated alike. All the relativization results stated below are full relativizations.

The first relativization result pertaining to the isomorphism conjecture was due to Kurtz [Kur83] who constructed an oracle relative to which the isomorphism conjecture failed. Recently this result was improved to

Theorem 6.7.1 ([Kur88]) *Relative to a generic oracle,[26] the complete m-degree of* NP *is made up of multiple 1-degrees.*

Building on the ideas in [Kur83], Hartmanis and Hemachandra showed

Theorem 6.7.2 ([HH87]) *There is an oracle relative to which* P = UP *and the complete m-degree of* NP *consists of multiple 1-degrees.*

This is a charming result in that it provides a relativized world in which *both* the isomorphism and the encrypted complete set conjectures fail. The constructions for both Theorem 6.7.1 and 6.7.2 make the isomorphism conjecture fail by building an oracle relative to which there is a "gappy" NP-complete set A. By this we mean that there are infinitely many n such that $\{x \in A : n \leq |x| \leq 2^n\}$ is empty. The existence (in the unrelativized world) of such a gappy NP-complete set runs counter to commonly held intuitions.

There is evidence that sparse oracles do not distort relationships among complexity classes. For example, it has been shown that the unrelativized polynomial-time hierarchy collapses if and only if there exists a sparse oracle relative to which the polynomial-time hierarchy collapses [LS86] [BBS86]. However, there is no evidence as to whether sparse oracle relativizations preserve fine details like the structure of the complete m-degree for NP. Be that as it may, the proofs of Theorems 6.5.2, 6.6.1, 6.6.2, and 6.6.6 can be modified to obtain

Theorem 6.7.3 ([KMR87b] [Lon88]) *There exists a sparse oracle relative to which the following holds:*

- *the complete m-degree for* NP *is a 1-li-degree;*

- *there is a collapsing degree that is 2-tt-complete for* NP;

- *there is a noncollapsing 1-li-degree that is 2-tt-complete for* NP; *and*

- *there is an m-degree that is 2-tt-complete for* NP *and that is made up of multiple 1-degrees.*

Goldsmith [Gol88] has related results obtained by somewhat different techniques.

If one dropped the sparsity condition on the oracle in the above theorem, then the result can be obtained trivially by choosing an oracle that makes NP = EXP, in which case Theorems 6.5.2, 6.6.1, 6.6.2, and 6.6.6 would apply at NP. However, it follows from [LS86, Theorem 3.2] that relative to a sparse oracle, NP = EXP if and only if in the unrelativized case NP = EXP.

---

[26] See [Rog67] and [Joc80] for a discussion of generic sets.

Until recently there was no known complexity class $\mathcal{C}$ for which there were oracles A and B such that (i) relative to A the complete m-degree for $\mathcal{C}$ collapsed and (ii) relative to B the complete m-degree for $\mathcal{C}$ failed to collapse. The breakthrough on this problem was made by Homer and Selman.

Theorem 6.7.4 ([HS89]) *(a) There exists an oracle relative to which the complete m-degree of Σ_2^P collapses.*

(b) There exists an oracle relative to which the complete m-degree of Σ_2^P fails to collapse.

The proof of part (b) is based on Kurtz's proof of an early version of Theorem 6.7.1. To show part (a), Homer and Selman cleverly construct an oracle A relative to which $\Sigma_2^P = \text{EXP}$ and $P = UP$, and thus, by Theorems 6.4.2 and 6.5.2, relative to A the complete m-degree of Σ_2^P collapses.

Shortly after Homer and Selman established Theorem 6.7.4, we showed

Theorem 6.7.5 ([KMR89]) *Relative to a random oracle[27] the complete 1-li-degrees of each of* NP, PSPACE, EXP, NEXP, *and* RE *all fail to collapse.*

To establish the theorem, we show that scrambling functions exist relative to random oracles and then apply Theorem 6.5.10 above. The proof that scrambling functions exist relative to random oracles involves an excursion into measure theory which we spare the reader in this survey. See [KMR89] for full details.

Putting together Proposition 6.5.9 and Theorems 6.7.4 and 6.7.5 we obtain

Corollary 6.7.6 *(a) There is an oracle relative to which the complete m-degrees of each of* NP, PSPACE, EXP, NEXP, *and* RE *all fail to collapse.*

(b) For each of PSPACE, EXP, NEXP, *and* RE, *there is an oracle relative to which the complete m-degree of the class collapses.*

---

[27]We say that a relativized statement T^X *holds measure one* if and only if the set $\{R : T^R\}$ has measure one. We say a set is *random* if and only if it *satisfies all arithmetically definable properties of measure one*. Thus, to show an arithmetically definable property holds relative to a random oracle, it suffices to show that the property holds measure one. Computability relative to a random oracle (very) roughly models computability under the hypothesis that strong, polynomial-time computable, pseudo-random functions exist. If one believes that such functions do exist, then random oracle results may be indicative of what is true about unrelativized computability.

Note that the notion of random set defined above is distinct from Chaitin's which is based on *algorithmic incompressibility*. There are Chaitin-random sets that are not random in the sense defined above.

Proof Sketch Part (a) follows directly from Theorem 6.7.5. The PSPACE portion of part (b) follows from Homer and Selman's proof of Theorem 6.7.4(a). The rest of part (b) follows from Proposition 6.5.9 and the existence of an oracle that makes P = NP.   □

An oracle that is notable for its absence in Corollary 6.7.6 is one that makes the original Berman and Hartmanis isomorphism conjecture true. The question of whether there is such an oracle is open and seems very difficult. When it was originally put forth, the isomorphism conjecture embodied some of the clearest insights into the structure of the NP-complete sets. It is ironic, then, that it seems so hard to obtain even a relativized confirmation of the conjecture.

Acknowledgements: Eric Allender made a number of very helpful suggestions on the first draft of this paper. Paul Young called us to task for an earlier inaccurate title. We would also like to thank Alan Selman for sharing with us his yet-unpublished research on the Joseph-Young conjecture.

Most of the work on this paper was done while the second author was at AT&T Bell Laboratories, Murray Hill and the third author was at the University of Chicago. The first author was supported in part by NSF Grant DCR-8602562. The third author was supported in part by NSF Grants DCR-8602991 and CCR-89011154.

6.8 REFERENCES

[All88]   E. Allender. Isomorphisms and 1-L reductions. *Journal of Computer and System Sciences*, 36:336–350, 1988.

[BBS86]   J. Balcázar, R. Book, and U. Schöning. The polynomial-time hierarchy and sparse oracles. *Journal of the ACM*, 33:603–617, 1986.

[Ben89]   C. Bennett. Time space tradeoffs for reversible computation. *SIAM Journal on Computing*, 1989. To appear.

[Ber77]   L. Berman. *Polynomial Reducibilities and Complete Sets*. PhD thesis, Cornell University, 1977.

[Ber78]   P. Berman. Relationship between density and deterministic complexity of NP-complete languages. In *Proceedings of the 5th International Colloquium on Automata, Languages, and Programming*, pages 63–71, Springer-Verlag, 1978. Lecture Notes in Computer Science No. 62.

[BGS75]   T. Baker, J. Gill, and R. Solovay. Relativizations of the P =? NP question. *SIAM Journal on Computing*, 4:431–442, 1975.

[BH77]   L. Berman and J. Hartmanis. On isomorphism and density of NP and other complete sets. *SIAM Journal on Computing*, 1:305–322, 1977.

[Can55]   G. Cantor. *Contributions to the Founding of the Theory of Transfinite Numbers*. Dover Publications, 1955.

[Cut80]   N. Cutland. *Computability: An Introduction to Recursive Function Theory*. Cambridge University Press, 1980.

[Dow78]   M. Dowd. On isomorphism. 1978. Unpublished manuscript.

[Dow82]   M. Dowd. *Isomorphism of Complete Sets*. Technical Report LCSR–TR–34, Laboratory for Computer Science Research, Busch Campus, Rutgers University, 1982.

[DW83]   M. Davis and E. Weyuker. *Computability, Complexity, and Languages*. Academic Press, 1983.

[Fen89]   S. Fenner. *A Complexity Theoretic Failure of the Cantor–Bernstein Theorem*. Technical Report 89–007, Department of Computer Science, University of Chicago, 1989.

[FKR89]   S. Fenner, S. Kurtz, and J. Royer. Every polynomial-time 1-degree collapses iff P = PSPACE. In *Proceedings of the 30th Annual IEEE Symposium on Foundations of Computer Science*, pages 624–629, 1989.

[For79]   S. Fortune. A note on sparse complete sets. *SIAM Journal on Computing*, 431–433, 1979.

[Gan89]   K. Ganesan. *Complete Problems, Creative Sets and Isomorphism Conjectures*. PhD thesis, Boston University, 1989.

[GH89]   K. Ganesan and S. Homer. Complete problems and strong polynomial reducibilities. In *Proceedings of STACS '89*, pages 240–250, 1989. Lecture Notes in Computer Science No. 349.

[GHS89]   J. Geske, D. Huyhn, and J. Seiferas. A note on almost-everywhere-complex sets and separating deterministic-time-complexity classes. *Information and Computation*, 1989. To appear.

[GJ79]   M. Garey and D. Johnson. *Computers and Intractability*. W. H. Freeman and Company, 1979.

[GJ86]   J. Goldsmith and D. Joseph. Three results on the polynomial isomorphism of complete sets. In *Proceedings of the 27th Annual IEEE Symposium on Foundations of Computer Science*, pages 390–397, 1986.

[Gol88]   J. Goldsmith. *Polynomial Isomorphisms and Near-Testable Sets*. PhD thesis, University of Wisconsin at Madison, 1988. Available as: Technical Report Number 816, Computer Sciences Department, University of Wisconsin at Madison.

[GS84]   J. Grollmann and A. Selman. Complexity measures for public-key cryptosystems. In *Proceedings of the 25th Annual IEEE Symposium on Foundations of Computer Science*, pages 495–503, 1984.

[GS88]   J. Grollmann and A. Selman. Complexity measures for public-key cryptosystems. *SIAM Journal on Computing*, 17:309–335, 1988.

[Har78a]   J. Hartmanis. *Feasible Computations and Provable Complexity Properties*. Society for Industrial and Applied Mathematics, 1978.

[Har78b]   J. Hartmanis. On log-tape isomorphisms of complete sets. *Theoretical Computer Science*, 273–286, 1978.

[HH87]   J. Hartmanis and L. Hemachandra. One-way functions, robustness, and the non-isomorphism of NP-complete sets. In *Proceedings of the 2nd Annual IEEE Structure in Complexity Theory Conference*, 1987.

[HIM81]   J. Hartmanis, N. Immerman, and S. Mahaney. One-way log-tape reductions. In *Proceedings of the 19th Annual IEEE Symposium on Foundations of Computer Science*, pages 65–72, 1981.

[HS89]   S. Homer and A. Selman. Oracles for structural properties. In *Proceedings of the 4th Annual IEEE Structure in Complexity Theory Conference*, pages 3–14, 1989.

[HU79]   J. Hopcroft and J. Ullman. *Introduction to Automata Theory, Languages, and Computation*. Addison-Wesley, 1979.

[Joc80]   C. Jockusch, Jr. Degrees of generic sets. In F. R. Drake and S. S. Wainer, editors, *Recursion Theory: Its Generalizations and Applications*, pages 110–139, Cambridge University Press, 1980.

[JY85]   D. Joseph and P. Young. Some remarks on witness functions for polynomial reducibilities in NP. *Theoretical Computer Science*, 39:225–237, 1985.

[KLD87]   K. Ko, T. Long, and D. Du. A note on one-way functions and polynomial-time isomorphisms. *Theoretical Computer Science*, 47:263–276, 1987.

[KMR87a]   S. Kurtz, S. Mahaney, and J. Royer. *Noncollapsing Degrees*. Technical Report 87-001, Department of Computer Science, University of Chicago, 1987.

[KMR87b]   S. Kurtz, S. Mahaney, and J. Royer. Progress on collapsing degrees. In *Proceedings of the 2nd Annual IEEE Structure in Complexity Theory Conference*, pages 126–131, 1987.

[KMR88]   S. Kurtz, S. Mahaney, and J. Royer. Collapsing degrees. *Journal of Computer and System Sciences*, 37:247–268, 1988.

[KMR89]   S. Kurtz, S. Mahaney, and J. Royer. The isomorphism conjecture fails relative to a random oracle. In *Proceedings of the 21st annual ACM Symposium on Theory of Computing*, pages 157–166, 1989.

[Ko85]   K. Ko. On some natural complete operators. *Theoretical Computer Science*, 37:1–30, 1985.

[Koz80]   D. Kozen. Indexings of subrecursive classes. *Theoretical Computer Science*, 11:277–301, 1980.

[Kur83]   S. Kurtz. A relativized failure of the Berman-Hartmanis conjecture. 1983. Unpublished manuscript.

[Kur88]   S. Kurtz. *The Isomorphism Conjecture Fails Relative to a Generic Oracle*. Technical Report 88-018, Department of Computer Science, University of Chicago, 1988.

[Lon88]   T. Long. One-way functions, isomorphisms, and complete sets. *Abstracts of the AMS*, 9:125, 1988.

[LS86]   T. Long and A. Selman. Relativizing complexity classes with sparse oracles. *Journal of the ACM*, 33:618–627, 1986.

[Mah81]   S. Mahaney. On the number of p-isomorphism classes of NP-complete sets. In *Proceedings of the 22th Annual IEEE Symposium on Foundations of Computer Science*, pages 271–278, 1981.

[Mah82]   S. Mahaney. Sparse complete sets for NP: Solution of a conjecture of Berman and Hartmanis. *Journal of Computer and System Sciences*, 25:130–143, 1982.

[Mah89]   S. Mahaney. The isomorphism conjecture and sparse sets. In J. Hartmanis, editor, *Computational Complexity Theory*, American Mathematical Society, 1989.

[Moo82]   G. Moore. *Zermelo's Axiom of Choice: Its Origins, Development, and Influence*. Springer-Verlag, 1982.

[MS72]   A. Meyer and L. Stockmeyer. The equivalence problem for regular expressions with squaring requires exponential time. In *Proceedings of the 13th Annual IEEE Symposium on Switching and Automata Theory*, pages 125–129, 1972.

[MWY78]   M. Machtey, K. Winklmann, and P. Young. Simple Gödel numberings. *SIAM Journal on Computing*, 7:39–60, 1978.

[MY78]   M. Machtey and P. Young. *An Introduction to the General Theory of Algorithms*. North-Holland, 1978.

[MY85]   S. Mahaney and P. Young. Reductions among polynomial isomorphism types. *Theoretical Computer Science*, 39:207–224, 1985.

[Myh55]   J. Myhill. Creative sets. *Zeitschrift für Mathematische Logik und Grundlagen der Mathematik*, 1:97–108, 1955.

[Myh59]   J. Myhill. Recursive digraphs, splinters and cylinders. *Mathematische Annalen*, 138:211–218, 1959.

[Pos44]   E. Post. Recursively enumerable sets of positive integers and their decision problems. *Bulletin of the AMS*, 50:284–316, 1944.

[RC87]   J. Royer and J. Case. *Intensional Subrecursion and Complexity Theory*. Technical Report 87–007, Department of Computer Science, University of Chicago, 1987.

[Rog58]   H. Rogers. Gödel numberings of the partial recursive functions. *Journal of Symbolic Logic*, 23:331–341, 1958.

[Rog67]   H. Rogers. *Theory of Recursive Functions and Effective Computability*. McGraw-Hill, 1967. Reprinted. MIT Press. 1987.

[Rus86]   D. Russo. Optimal approximations of complete sets. In *Proceedings of the Structure in Complexity Theory Conference*, pages 311–324, Springer-Verlag, 1986.

[Sch75]   C. Schnorr. Optimal enumerations and optimal Gödel number-
           ings. *Mathematical Systems Theory*, 8:182–191, 1975.

[Wat85]   O. Watanabe. On one-one polynomial time equivalence rela-
           tions. *Theoretical Computer Science*, 38:157–165, 1985.

[Wat88]   O. Watanabe. A note on p-isomorphism conjecture. 1988. Un-
           published manuscript.

[You66]   P. Young. Linear orderings under one-one reducibility. *Journal
           of Symbolic Logic*, 31:70–85, 1966.

7

Applications of Kolmogorov Complexity in the Theory of Computation

Ming Li[1]
Paul M.B. Vitányi[2]

ABSTRACT This exposition gives a brief introduction to the main ideas of
Kolmogorov complexity that have been useful in the area of computational
complexity theory. We demonstrate how these ideas can actually be applied
and provide a detailed survey of the abundant applications of this elegant
notion in computational complexity theory. (*Note* : Preliminary versions of
parts of this paper appeared in: *Proc. 3rd IEEE Structure in Complexity
Theory Conference* , Computer Society Press, Washington D.C., 1988, pp.
80-102; and *Uspekhi Mat. Nauk*, 43:6 (1988), pp. 129-166 (in Russian).)

7.1 Introduction

The theory of computation is primarily concerned with the analysis and
synthesis of algorithms in relation to the resources in time and space such
algorithms require. R.J. Solomonoff [Sol64], A.N. Kolmogorov [Kol65], and
G.J. Chaitin [Cha69] (in chronological order) have invented an excellent
theory, now commonly called Kolmogorov complexity, of information content of strings that is most useful for this pursuit. Intuitively, the amount
of information in a finite string is the size (literally, number of bits) of the
shortest program that, starting with a blank input, computes the string and
then terminates. A similar definition can be given for infinite strings, but

[1]Department of Computer Science, York University, Ontario M3J 1P3,
Canada. The work of the first author was partially performed at Aiken Com-
putation Lab, Harvard University and supported by NSF Grant DCR-8606366,
Office of Naval Research Grant N00014-85-k-0445, Army Research Office Grant
DAAL03-86-K-0171, and NSERC Operating Grant OGP0036747. Current ad-
dress: Computer Science Department, University of Waterloo, Waterloo, Ontario,
Canada N2L 3G1. (mli@water.waterloo.edu).

[2]Centrum voor Wiskunde en Informatica and Faculteit Wiskunde en Informatica, Universiteit van Amsterdam, Kruislaan 413, 1098 SJ Amsterdam, The
Netherlands (paulv@piring.cwi.nl).

in this case the program produces element after element forever. Thus, 1^n (a string of n 1's) contains little information because the following program of size $O(\log n)$ outputs it:

$$\text{for } i := 1 \text{ to } n \text{ do print('1').}$$

Likewise, the transcendental number $\pi = 3.1415\ldots$, an infinite sequence of seemingly "random" decimal digits, contains $O(1)$ information. (There is a short program that produces the consecutive digits of π forever.) Such a definition would appear to make the amount of information in a string depend on the particular programming language used. Fortunately, it can be shown that all choices of programming languages (that make sense) lead to quantification of the amount of information that is invariant up to an additive constant.

In the past ten years, Kolmogorov complexity has been successfully applied to many areas including computational complexity. Although Kolmogorov complexity contains rich and deep mathematics, the amount of this mathematics one needs to know to apply the notion fruitfully in the area of computational complexity is little. However, formal knowledge does not necessarily imply the wherewithal to apply it, perhaps especially so in the case of Kolmogorov complexity. It is the purpose of this exposition to develop the minimum amount of theory needed, and outline a scala of illustrative applications in the study of structure in complexity theory. In fact, while the pure theory of the subject will have appeal to a select few, a surprisingly large amount of its applications will, we hope, delight the multitude.

7.2   A Simplified Mathematical Theory

In order to give a simplest but sufficient treatment of Kolmogorov complexity for the applications, we will deal with only finite strings. We will also stick with the most basic type of Kolmogorov complexity and discuss no other variants. Although the mathematical theory of Kolmogorov complexity extends far beyond our treatment given here, we will treat only the essential ideas and useful facts that are used in the applications to come. We refer those readers who are interested in a more complete treatment of Kolmogorov complexity or a detailed survey of applications of Kolmogorov complexity to [LVb], [LV88a] or [LV89a], and to our forthcoming book [LVa] where one can also find most of the proofs that are missing here. The classic early survey in the area is Levin and Zvonkin's treatment [ZL70].

We are interested in defining the complexity of a concrete individual finite string of zeros and ones. Unless otherwise specified, all strings will be binary and of finite length. All logarithms in this paper are base 2, unless it is explicitly noted they are not. If x is a string, then $|x|$ denotes the *length* (number of zeros and ones) of x. We identify throughout the xth finite binary string with the natural number x, according to the correspondence:

$$(\epsilon, 0), (0, 1), (1, 2), (00, 3), (01, 4), (10, 5), (11, 6), (000, 7), \ldots$$

If A is a set, then $d(A)$ is the *cardinality* (the number of elements) of A. Intuitively, we want to call a string simple if it can be described in a few words, like "the string of a million ones". A string is considered complex if it cannot be so easily described, for example a "random" string does not follow any rule and hence we do not know how to describe it apart from giving it literally. A description of a string may depend on two things, the decoding method (the machine which interprets the description) and outside information available (input to the machine). We are interested in descriptions which are effective, and restrict the decoders to Turing machines. Without loss of generality, our Turing machines use binary input strings which we call *programs* . More formally, fixing a Turing machine T, we would like to say that p is a description of x if, on input p, T outputs x. It is also convenient to allow T to have some extra information y to help to generate x. We write $T(p, y) = x$ to mean that Turing machine T with input p and y terminates with output x.

Definition 7.2.1 *The descriptional complexity K_T of x, relative to Turing machine T and binary string y, is defined by*

$$K_T(x|y) = min\{|p| : p \in \{0,1\}^* \ \& \ T(p, y) = x\}.$$

The complexity measure defined above is useful and makes sense only if the complexity of a string does not depend on the choice of T. Therefore the following simple theorem is vital.

Theorem 7.2.2 (Invariance Theorem [Sol64,Kol65,Cha69])
There exists a (optimal) Turing machine U, such that, for any other Turing machine T, there is a constant c_T such that for all strings x, y, $K_U(x|y) \le K_T(x|y) + c_T$.

Proof. Fix some standard enumeration of Turing machines $T_1, T_2, \ldots$. Let U be the Universal Turing machine such that when starting on input $0^n 1p$, $p \in \{0,1\}^*$, U simulates the nth Turing machine T_n on input p. For convenience in the proof, we choose U such that if T_n halts, then U first erases everything apart from the halting contents of T_n's tape, and also halts. By construction, for each $p \in \{0,1\}^*$, T_n started on p eventually halts if and only if U started on $0^n 1p$ eventually halts. Choosing $c_T = n+1$ for T_n finishes the proof.   □

Clearly, the Universal Turing machine U that satisfies the Invariance Theorem is *optimal* in the sense that K_U minorizes each K_T up to a fixed additive constant (depending on U and T). Moreover, for each pair of Universal Turing machines U and U', satisfying the Invariance Theorem, the complexities coincide up to an additive constant (depending only on U and U'), for all strings x, y:

$$|K_U(x|y) - K_{U'}(x|y)| \leq c_{U,U'}.$$

Therefore, we set the canonical *conditional Kolmogorov* complexity $K(x|y)$ of x under condition of y equal to $K_U(x|y)$, for some fixed optimal U. We call U the *reference* Turing machine. Hence the Kolmogorov complexity of a string does not depend on the choice of encoding method and is well-defined. Define the *unconditional Kolmogorov* complexity of x as $K(x) = K(x|\epsilon)$, where ϵ denotes the empty string ($|\epsilon| = 0$).

Definition 7.2.3 *A binary string x is* incompressible *if* $K(x) \geq |x|$.

Remark. Since Martin-Lof has shown that incompressible strings pass all effective statistical tests for randomness, we will also call incompressible strings *random* strings.

We now start to prove several useful and simple properties. We name those properties that will be needed for later applications as Facts, and some other properties, that are just "nice" to know, are listed as Examples.

Example 7.2.4 For each finite binary string x we have $K(xx) \leq K(x) + O(1)$. Namely, let T compute x from program p. Let $n(T)$ be the position number of T in the effective enumeration of all TM's. Now fix a universal machine V which, on input $0^{n(T)}1p$, simulates T just like the reference machine U in the proof of the Invariance Theorem, but additionally V doubles T's output before halting. Now V started on $0^{n(T)}1p$ computes xx, and therefore U started on $0^{n(V)}10^{n(T)}1p$ computes xx. Hence, for all x, $K(xx) \leq K(x) + n(V) + 1$.

Example 7.2.5 Let us define $K(x, y) = K(<x, y>)$ with $< \cdot, \cdot >$ is a standard one-one mapping of pairs of natural numbers to natural numbers. That is, $K(x, y)$ is the length of a shortest program that outputs x and y and a way to tell them apart. It is seductive to conjecture $K(x, y) \leq K(x) + K(y) + O(1)$, the obvious (but false) argument running as follows. Suppose we have a shortest program p to produce x, and a shortest program q to produce y. Then with $O(1)$ extra bits to account for some Turing machine T that schedules the two programs, we have a program to produce x followed by y with a separator in between. However, any such T will have to know where to divide its input to identify p and q. We can separate p and q by prefixing pq by a clearly distinguishable encoding r of the length $|p|$ in $O(\log |p|)$ bits (see next Section on self-delimiting strings). Consequently, we have at best established $K(x, y) \leq K(x) + K(y) + O(\log(min(K(x), K(y))))$. In general this cannot be improved.

Fact 7.2.6 *For each n, there exists a random string of length n.*

Proof. Since there are 2^n binary strings of length n, but only $2^n - 1$ possible shorter descriptions, it follows that, for all n, there is a binary string x of length n such that $K(x) \geq n$. □

Fact 7.2.7 *If $x = uvw$, then $K(v) \geq K(x) - |uw| - O(\log|x|)$. Hence if x is random; i.e., $K(x) \geq |x|$, then $K(v) \geq |v| - O(\log|x|)$.*

Proof. A string $x = uvw$ can be specified by a description of v, literal descriptions of the binary representation of $|u|$ and the concatenation uw. Additionally, we need information to tell these three items apart. Such information can be provided in $O(\log|x|)$ bits. Thus,

$$K(x) \leq K(v) + O(\log|x|) + |uw|.$$

Rearranging above, for random string x with $K(x) \geq |x|$ we obtain

$$K(v) \geq |v| - O(\log|x|).$$

□

This fact shows that all long substrings of a random string are also almost incompressible. It can be shown that this is optimal, a substring of an incompressible string may be compressible by $O(\log|x|)$ bits. This conforms to a fact we know from probability theory: every sufficiently long random string must contain long runs of zeros.

Fact 7.2.8 *If, for sufficiently long x, $K(x) \geq |x|$, then no substring of length longer than $3\log|x|$ occurs in x more than once (i.e., $u_1vw_1 = uvw$ for $u_1 \neq u$).*

Proof. Let $x = uvw$, $|x| = n$, and v of length $3\lceil \log n \rceil$ occurs twice in uv. To describe x, we only need to concatenate the following information, all items encoded so that we can tell them apart (cf. also Example 7.2.14):

- this discussion;

- the location of the beginning of the first instance of v in uv, using at most $\lceil \log n \rceil + 2\log\log n$ bits;

- the location of the beginning of the second instance of v in uv, using at most $\lceil \log n \rceil + 2\log\log n$ bits;

- the length of v in at most $2\log\log n$ bits;

- the length of u in at most $2 \log \log n$ bits;

- the literal word uw, using exactly $|uw|$ bits.

(This is a description of uvw, since $u = u'v_1$ with $v = v_1 v'$, so $v = v_1 v_1 v''$; repeatedly exploiting this idea allows us to recover all of v.) Altogether this description requires at most $n - \log n + 8 \log \log n + O(1)$ bits. For n large enough we have $K(x) < n$, a contradiction.   □

Example 7.2.9 Apart from showing that complexity is an attribute of the finite object alone, the Invariance Theorem also has another most important consequence: it gives an upper bound on the complexity. Namely, there is a fixed constant c such that for all x of length n we have

$$K(x) \le n + c.$$

This is easy to see. If T is a machine that just copies its input to its output, then $p = 0^{n(T)} 1 x$ is a program for the reference machine U to output x. This says that $K(x)$ is bounded above by the length of x modulo an additive constant.

Example 7.2.10 Define $p(x)$ as a shortest program for x. We show that $p(x)$ is incompressible. There is a constant $c > 0$, such that for all strings x, we have $K(p(x)) \ge c \cdot |p(x)|$. For suppose the contrary, and there is a program $p(p(x))$ that generates $p(x)$ with $|p(p(x))| \le c \cdot |p(x)|$. Define a universal machine V that works just like the reference machine U, except that V first simulates U on its input to obtain an output, and then uses this output as input on which to simulate U once more. But then, U with input $0^{n(V)} 1 p(p(x))$ computes x, and therefore $K(x) \le c \cdot |p(x)| + n(V) + 1$. However, this yields $(1 - c)K(x) \le n(V) + 1$, for all x, which is impossible by a trivial counting argument. Similarly we can show that there is a $c > 0$ such that for all strings x, we have $K(p(x)) \ge |p(x)| - c$.

Example 7.2.11 It is easy to see that $K(x|x) \le |n(T)| + 1$, where T is a machine that just copies the input to the output. However, it is more interesting that $K(p(x)|x) \le \log K(x) + O(1)$, which cannot be improved in general. (Hint: later we show that K is a noncomputable function.) This rules out that we can compute $p(x)$ from x. However, we can dovetail the computation of all programs shorter than $|x| + 1$: run the 1st program 1 step, run the 1st program 1 step and 2nd program 1 step, and so on. This way we will eventually enumerate all programs that output x. However, since some computations may not halt, and the halting problem is undecidable, we need to know the length of a shortest program $p(x)$ to recognize such a program when it is found.

A natural question to ask is: how many strings are incompressible? It turns out that virtually all strings of given length n are incompressible. Namely, there is at least one x of length n that cannot be compressed to length $< n$ since there are 2^n strings of length n and but $2^n - 1$ programs of length less than n; at least $1/2$ of all strings of length n cannot be compressed to length $< n - 1$ since there are but $2^{n-1} - 1$ programs of length less than $n - 1$; at least $3/4$th of all strings of length n cannot be compressed to length $< n - 2$, and so on.

Generally, let $g(n)$ be an integer function. Call a string x of length n, g-incompressible if $K(x) \geq n - g(n)$. There are 2^n binary strings of length n, and only $2^{n-g(n)} - 1$ possible descriptions shorter than $n - g(n)$. Thus, the ratio between the number of strings x of length n with $K(x) < n - g(n)$ and the total number of strings of length n is at most $2^{-g(n)+1}$, a *vanishing fraction* when $g(n)$ increases unboundedly with n. In general we loosely call a finite string x of length n *random* if $K(x) \geq n - O(\log n)$.

Intuitively, incompressibility implies the absence of regularities, since regularities can be used to compress descriptions. Accordingly, we like to identify incompressibility with absence of regularities or *randomness* . In the context of finite strings randomness like incompressibility is a matter of degree: it is obviously absurd to call a given string random and call nonrandom the string resulting from changing a bit in the string to its opposite value. Thus, we identify c-incompressible strings with *c-random* strings.

Fact 7.2.12 *Fix any consistent and sound formal system F. For all but finitely many random strings x, the sentence "x is random" is not provable in F.*

Proof. Assume that the fact is not true. Then given F, we can start to exhaustively search for proof that some string of length $n \gg |F|$ is random, and print it once we find such a string x. This procedure to print x of length n is only $O(\log n) + |F| \ll n$. But x is random by the proof. Hence F is not consistent. $\square$

The above fact shows that though most strings are random, it is impossible to effectively prove them random. This in a way explains why Kolmogorov complexity is so successful in many applications when effective construction of a required string is difficult by other methods. The fact that almost all finite strings are random but cannot be proved to be random amounts to an information-theoretic version of Gödel's theorem and follows from the noncomputability of $K(x)$. Strings that are not incompressible are *compressible* or *nonrandom* .

Fact 7.2.13 (Kolmogorov)

(a) *The function $K(x)$ is not partial recursive. Moreover, no partial recursive function $\phi(x)$, defined on an infinite set of points, can coincide with $K(x)$ over the whole of its domain of definition. (Every unbounded partial recursive function "over-estimates" $K(x)$ infinitely often (on its domain).)*

(b) *There is a (total) recursive function $H(t, x)$, monotonically nonincreasing in t, such that $\lim_{t \to \infty} H(t, x) = K(x)$. That is, we can obtain arbitrary good estimates for $K(x)$ (but not uniformly).*

Proof.

(a) Every infinite r.e. set contains an infinite recursive subset, Theorem 5-IV in [Rog67]. Select an infinite recursive set A in the domain of definition of $\phi(x)$. The function $f(m) = min\{x : K(x) \geq m, x \in A\}$ is (total) recursive (since $K(x) = \phi(x)$ on A), and takes arbitrary large values. Also, by construction $K(f(m)) \geq m$. On the other hand, $K(f(m)) \leq K_f(f(m)) + c_f$ by definition of K, and obviously $K_f(f(m)) \leq |m|$. Hence, $m \leq \log m$ up to a constant independent of m, which is false.

(b) Let c be a constant such that $K(x) \leq |x| + c$ for all x. Define $H(t, x)$ as the length of the smallest program p, with $|p| \leq |x| + c$, such that the reference machine U with input p halts with output x within t steps.

$\square$

Example 7.2.14 (Self-delimiting Descriptions)

Let the variables x, y, x_i, y_i ... denote strings in $\{0, 1\}^*$. A *description* of x, $|x| = n$, can be given as follows.

(1) A piece of text containing several formal parameters $p_1, \ldots, p_m$. Think of this piece of text as a formal parametrized procedure in an algorithmic language like C. It is followed by

(2) an ordered list of the actual values of the parameters.

The piece of text of (1) can be thought of as being encoded over a given finite alphabet, each symbol of which is coded in bits. Therefore, the encoding of (1) as prefix of the binary description of x requires $O(1)$ bits. This prefix is followed by the ordered list (2) of the actual values of $p_1, \ldots, p_m$ in binary. To distinguish one from the other, we encode (1) and the different items in (2) as self-delimiting strings, an idea used already by C.E. Shannon.

For each string $x \in \{0,1\}^*$, the string $\bar{x}$ is obtained by inserting a "0" in between each pair of adjacent letters in x, and adding a "1" at the end. That is,

$$\overline{01011} = 0010001011.$$

Let $x' = \overline{|x|}x$ (the length of x in binary followed by x in binary). The string x' is called the *self-delimiting* version of x. So '100101011' is the self-delimiting version of '01011'. (Note that according to our convention "10" is the 5th binary string.) The self-delimiting binary version of a positive integer n requires $\log n + 2 \log \log n$ bits, and the self-delimiting version of a binary string w requires $|w| + 2 \log |w|$ bits. For convenience, we denote the length $|n|$ of a natural number n by "$\log n$".

Example 7.2.15 Self-delimiting descriptions were used in the proof of the Invariance Theorem. (Namely, in the encoding $0^{n(T)}1$.) Using it explicitly, we can define Kolmogorov complexity as follows. Fix an effective coding C of all Turing machines as binary strings such that no code is a prefix of any other code. Denote the code of Turing machine M by $c(M)$. Then the Kolmogorov complexity of $x \in \{0,1\}^*$, with respect to c, is defined by $K_c(x) = min\{|c(M)y| : M \text{ on input } y \text{ halts with output } x\}$.

Example 7.2.16 (Self-delimiting Kolmogorov complexity) A code c such that for all x, y, $c(x)$ is not a prefix of $c(y)$ if $x \neq y$ is called a *prefix code*. We can define a variant of Kolmogorov complexity by requiring at the outset that we only consider Turing machines for which the set of programs is a prefix code. The resulting variant, called *self-delimiting* Kolmogorov complexity, has nicer mathematical properties than the original one, and has therefore become something of a standard in the field. For our applications it does not matter which version we use since they coincide to within an additive term of $O(\log |x|)$.

Example 7.2.17 (M. Li, W. Maass, P.M.B. Vitányi) In proving lower bounds in the theory of computation it is sometimes useful to give an efficient description of an incompressible string with 'holes' in it. The reconstruction of the complete string is then achieved using an additional description. In such an application we aim for a contradiction where these two descriptions together have significantly smaller length than the incompressible string they describe. Formally, let $x = x_1 \ldots x_k$ be a binary string of length n with the x_i's ($1 \leq i \leq k$) blocks of equal length c. Suppose that d of these blocks are deleted and the relative distances in between deleted blocks are known. We can describe this information by: (1) a formalization of this discussion in $O(1)$ bits, and (2) the actual values of

$$c, m, p_1, d_1, p_2, d_2, \ldots, p_m, d_m,$$

where m ($m \leq d$) is the number of "holes" in the string, and (3) the literal representation of

$$\hat{x} = \hat{x}_1 \hat{x}_2 \ldots \hat{x}_k.$$

Here $\hat{x}_i$ is x_i if it is not deleted, and is the empty string otherwise; p_j, d_j indicates that the next p_j consecutive x_i's (of length c each) are one contiguous group followed by a gap of $d_j c$ bits long. Therefore, $k - d$ is the number of (non-empty) $\hat{x}_i$'s, with

$$k = \sum_{i=1}^{m}(p_i + d_i) \ \ \& \ \ d = \sum_{i=1}^{m} d_i.$$

The actual values of the parameters and $\hat{x}$ are coded in a self-delimiting manner. Then, by the convexity of the logarithm function, the total number of bits needed to describe the above information is no more than:

$$(k - d)c + 3d\log(k/d) + O(\log n).$$

We then proceed by showing that we can describe x by this description plus some description of the deleted x_i's, so that the total requires considerably less than n bits. Choosing x such that $K(x) \geq n$ then gives a contradiction. See [LV88b] and [Maa85].

Example 7.2.18 (Quantitative Estimates) Consider the conditional complexity of a string x, with x an element of a given finite set M, given some string y. Let $d(M)$ denote the number of elements in M. Then the fraction of $x \in M$ for which $K(x|y) < |d(M)| - m$, does not exceed 2^{-m+1}. This can be shown by a counting argument similar to before. Hence we have shown that the conditional complexity of the majority of elements in a finite set cannot be significantly less than the logarithm of the size of that set. The following Lemma says that it can not be significantly more either.

Lemma 7.2.19 (Kolmogorov) *Let A be an r.e. set of pairs (x, y), and let $M_y = \{x : (x, y) \in A\}$. Then, up to a constant depending only on A, $K(x|y) \leq |d(M_y)|$.*

    Proof. Let A be enumerated by a Turing machine T. Using y, modify T to T_y such that T_y enumerates all pairs (x, y) in A, without repetition. In order of enumeration we select the pth pair (x, y), and output the first element; i.e., x. Then we find $p < d(M_y)$, such that $T_y(p) = x$. Therefore, we have by the Invariance Theorem $K(x|y) \leq K_{T_y}(x) \leq |d(M_y)|$, as required. $\qquad\square$

Example 7.2.20 Let A be a subset of $\{0, 1\}^*$. Let $A^{\leq n}$ equal $\{x \in A : |x| \leq n\}$. If the limit of $d(A^{\leq n})/2^n$ goes to zero as n goes to infinity, then we call A *sparse* . For example, the set of all finite strings that have twice as many zeros as ones $d(A^{\leq n})/2^n \approx 2^{0.0817n}$ is sparse. This has as a consequence that all but finitely many of these strings have short programs.

Claim 7.2.21 (a) *(M. Sipser) If A is recursive and sparse, then for all constant c there are only finitely many x in A with $K(x) \geq |x| - c$. Using Kolmogorov's Lemma we can extend this result as follows.*

(b) *If A is r.e. and $d(A^{\leq n}) = o(n^{-(1+\epsilon)}2^n)$, $\epsilon > 0$, then, for all constant c, there are only finitely many x in A with $K(x) \geq |x| - c$.*

(c) *If A is r.e. and $d(A^{\leq n}) \leq p(n)$ with p a polynomial, then, for all constant $c > 0$, there are only finitely many x in A with $K(x) \geq |x|/c$.*

Proof.

(a) Consider the lexicographic enumeration of all elements of A. There is a constant d, such that the ith element x of A has $K(x) \leq K(i) + d$. If x has length n, then the sparseness of A implies that $K(i) \leq n - g(n)$, with $g(n)$ unbounded. Therefore, for each constant c, and all n, if x in A is of length n then $K(x) < n - c$, from some n onward.

(b) Fix c. Consider an enumeration of n-length elements of A. For all such x, the Lemma 7.2.19 above in combination with the sparseness of A implies that $K(x|n) \leq n - (1 + \epsilon) \log n + O(1)$. Therefore, $K(x) \leq n - \epsilon \log n + O(1)$, for some other positive ϵ, and the right hand side of the inequality is less than $n - c$ from some n onward.

(c) Similar to above.

$\square$

Next we prove an important fact known as the Symmetry of Information Lemma, due to Kolmogorov and Levin.

Fact 7.2.22 (Symmetry of Information) *To within an additive term of $O(\log K(x, y))$,*

$$K(x, y) = K(x) + K(y|x).$$

In the general case it has been proved that equality up to a logarithmic error term is the best possible. From Fact 7.2.22 it follows immediately that, to within an additive term of $O(\log K(x, y))$,

$$K(x) - K(x|y) = K(y) - K(y|x).$$

Proof. ($\leq$) We can describe both x and y by giving a description of x, a description of y given x, and an indication of where to separate the two descriptions. If p is a shortest program for x, and q is a shortest program for y, then $\overline{p}pq$ is a description for (x, y). Thus, up to an additive constant term, $K(x, y) \leq K(x) + K(y|x) + 2\log(K(x))$. Obviously, $K(x) \leq K(x, y)$ up to a constant.

($\geq$) For convenience we show $K(y|x) \leq K(x, y) - K(x) + O(\log K(x, y))$.
Assume this is not true. Then, for every constant c there is an x and y
such that $K(y|x) \geq K(x, y) - K(x) + c\log(K(x, y))$. Consider a description
of y using its serial number in the recursively enumerable set $Y = \{z :
K(x, z) \leq K(x, y)\}$. Given x, we can describe y by its serial number and
a description of $K(x, y)$, so $K(y|x) \leq \log(d(Y)) + 2\log(K(x, y))$, up to a
fixed independent constant. By the contradictory assumption, $d(Y) > 2^d$
with $d = K(x, y) - K(x) + c\log(K(x, y))$. But now we can obtain a too
short description for x as follows.

Each distinct pair (u, z) with $K(u, z) \leq K(x, y)$ requires a distinct description of length at most $K(x, y)$. Since there are no more than $2^{K(x,y)+1}$
such descriptions, we can reconstruct x from its serial number j in a recursively enumerable set that is small. Namely, given $K(x, y)$ and d, consider
the strings u such that $d(\{z : K(u, z) \leq K(x, y)\}) > 2^d$. Denote the recursively enumerable set of such u by X. Clearly, $x \in X$. By the upper bound
on the available number of distinct descriptions, $d(X) < 2^{K(x,y)+1}/2^d$. We
can now reconstruct x from $\overline{K(x, y)d}j$. That is, up to an independent constant, $K(x) \leq 2\log(K(x, y)) + 2\log(d) + K(x, y) - d + 1$. For c large this
implies the contradiction $K(x) < K(x)$.   □

Fact 7.2.23 *All previous Facts are true relative to a given binary string*
y.

For example, Fact 7.2.6 can be rewritten as: For all finite strings y and all
n there is an x of length n such that $K(x|y) \geq |x|$. And the equation for
symmetry of information can be relativized with a string z:

$$K(x|z) - K(x|y|z) = K(y|z) - K(y|x|z).$$

We have introduced the basic theory of Kolmogorov complexity. We now
turn to its applications in the complexity theory.

7.3   Proving Lower Bounds

It was observed in [PSS81] that static, descriptional (program size) complexity of a *single* random string can be used to obtain lower bounds on dynamic, computational (running time) complexity. The power of the static,
descriptional Kolmogorov complexity in the dynamic, computational lower
bound proofs rests on one single idea: there are incompressible (or Kolmogorov random) strings. A traditional lower bound proof by counting,
usually involves *all* inputs (or all strings of a certain length) and one shows
that the lower bound has to hold for *some* of these ("typical") inputs.
Since a particular "typical" input is *hard* to construct, the proof has to

involve all the inputs. Now we understand that a "typical input" can be constructed via a Kolmogorov random string. However, as we have shown no such string can be proved to be random or "typical". In a Kolmogorov complexity proof, we choose a random string that *exists*. That it cannot be exhibited is no problem, since we only need existence. Routinely, the way one proves a lower bound by Kolmogorov complexity is as follows. Fix a Kolmogorov random string which we know exists. Prove the lower bound with respect to this particular *fixed* string: show that if the lower bound does not hold, then this string can be compressed. Because we are dealing with only one fixed string, the lower bound proof usually becomes quite easy and natural.

In the next sub-section, we give three examples to illustrate the basic methodology. In the following sub-sections, we survey the lower bound results obtained using Kolmogorov complexity over the past 10 years (1979-1988), and sometimes we also include proofs which are either unpublished and short or illustrative and short. Many of these results resolve old or new, some of them well-known, open questions; Some of these results greatly simplify and improve the existing proofs.

7.3.1 THREE EXAMPLES OF PROVING LOWER BOUNDS

In this section, we illustrate how Kolmogorov complexity is used to prove lower bounds by three concrete examples, each with increasingly more difficult corresponding counting arguments. Example 7.3.1 is rather trivial and it is presented for a reader who has never seen a proof using Kolmogorov complexity.

Example 7.3.1 (Converting NFA to DFA) It requires $\Omega(2^n)$ states to convert a nondeterministic finite automaton to a deterministic finite automaton.

Proof. Consider the language $L_k = \{x \mid$ from the right, the kth bit of x is $1\}$. L_k can be accepted by a NFA with $k+1$ states. Suppose a DFA A with only $o(2^k)$ states accepts L_k. We will also write A as its own description. Fix a string x of length k such that $K(x \mid A, k) \geq |x|$. Give $x0^*$ to A as input. Stop A when it reads the last bit of x. Record the current state of A. Reconstruct x by running A starting from the recorded current state; feed A with input 0's; at the $i-1$st 0, if A accepts, then the ith bit of x is 1, otherwise it is 0. This needs only $k - f(k)$ bits, $f(k)$ unbounded, since A has only $o(2^k)$ states. So for large enough k we have $K(x|A,k) < |x|$, which is a contradiction.   □

Example 7.3.2 (One Tape Turing Machines, Paul) Consider a most basic Turing machine model with only one tape, which serves as both input and work tape and with a two-way read/write head. The input is initially put on the first n cells of the tape. We refer a reader who is not familiar with Turing machines to [HU79] for a detailed definition. The following theorem was first proved by Hennie and a proof by counting, for comparison, can be found in [HU79] page 318. Paul [Pau79] presented the following elegant proof. Historically, this was the first Kolmogorov-complexity lower bound proof.

Theorem 7.3.3 *It requires $\Omega(n^2)$ steps for the above single tape TM to recognize $L = \{ww^R : w \in \{1,0\}^*\}$ (the palindromes).*

Proof [Pau79]. Assume on the contrary, M accepts L in $o(n^2)$ time. Let $|M|$ denote the length of the description of M. Fix a Kolmogorov random string w of length $n/3$ for a large enough n. Consider the computation of M on ww^R. A *crossing sequence* associated with a tape square consists of the sequence of states the finite control is in when the tape head crosses the intersquare boundary between this square and its left neighbor. If $c.s.$ is a crossing sequence, then $|c.s.|$ denotes the length of its description. Consider an input of $w0^{n/3}w^R$ of length n. Divide the tape segment containing the input into three equal length segments of size $n/3$. If each crossing sequence associated with a square in the middle segment is longer than $\frac{n}{10|M|}$ then M spent $\Omega(n^2)$ time on this input. Otherwise there is a crossing sequence of length less than $\frac{n}{10|M|}$. Assume that this occurs at c_0. Now this crossing sequence requires at most $n/10$ bits to encode. Using this crossing sequence, we re-construct w as follows. For every string $x0^{n/3}x^R$ of length n, put it on the input tape and start to simulate M. Each time the head reaches c_0 from the left, we take the next element in the crossing sequence to skip to the computation of M when the head is on the right of c_0 and resume the simulation starting from the time when the head moves back to the left of (or on) c_0 again. If the simulation ends consistently; i.e., every time the head moves to c_0 the current status of M is consistent with that specified in the crossing sequence, then $w = x$. Otherwise, if $w \neq x$ and the crossing sequences are consistent in both computations, then M accepts a wrong input $x0^{n/3}w^R$. However this implies

$$K(w) < |c.s.| + O(\log n) < n,$$

contradicting $K(w) \geq n$.   □

Example 7.3.4 (Boolean Matrix Rank, Seiferas-Yesha) For all n, there is an n by n matrix over $GF(2)$ (a matrix with zero-one entries with the usual boolean multiplication (and) and addition (xor)) such that every submatrix of s rows and $n - r$ columns ($r, s \leq n/4$) has at least $s/2$ linear independent rows.

Remark. Combined with the results in [BC80] and [Yes84] this example implies that $TS = \Omega(n^3)$ optimal lower bound for Boolean matrix multiplication on any general random access machines, where T stands for time and S stands for space.

Proof. Fix a random sequence x of elements in $GF(2)$ (zeros and ones) of length n^2, so $K(x) \geq |x|$. Arrange the bits of x into a matrix M, one bit per entry in, say, the row-major order. We claim that this matrix M satisfies the requirement. To prove this, suppose this is not true. Then consider a submatrix of M of s rows and $n - r$ columns, $r, s \leq n/4$. Suppose that there are at most $\frac{s}{2} - 1$ linearly independent rows. Then $1 + \frac{s}{2}$ rows can be expressed by the linear combination of other $\frac{s}{2} - 1$ rows. Thus we can describe this submatrix using

- the $\frac{s}{2} - 1$ linear independent rows, in $(\frac{s}{2} - 1)(n - r)$ bits;

- for each of the other $\frac{s}{2} + 1$ rows, use $(\frac{s}{2} - 1)2$ bits.

This saves $(\frac{s}{2} + 1)(n - r - \frac{s}{2} + 1)$ bits. Then to specify x, we only need to specify, in addition to the above, (a) M without the bits of the submatrix and (b) the indices of the columns and rows of this submatrix. When we list the indices of the rows of the submatrix, we list the $\frac{s}{2} - 1$ linearly independent rows first. Hence we only use

$$n^2 - (n-r)s + (n-r)\log n + s\log n + (\frac{s}{2} - 1)(n-r) + (\frac{s}{2} - 1)(\frac{s}{2} + 1) < n^2$$

bits, for large n's. This contradicts the fact $K(x) \geq |x|$.   □

Remark. A lower bound obtained by Kolmogorov complexity usually implies that the lower bound holds for "almost all strings". This is the case for all three examples. In this sense the lower bounds obtained by Kolmogorov complexity are often stronger than those obtained by its counting counterpart, since it usually also implies directly the lower bounds for nondeterministic or probabilistic versions of the considered machine.

7.3.2 ONE MORE EXAMPLE

In this section we present a more difficult proof which solved an open problem in computational complexity theory. The history of this problem, as well as the difference between the TM model and the one in Example 7.3.2, will be discussed in the following section. We actually present a weaker form of the theorem and suggest only very serious readers read this.

Theorem 7.3.5 *It requires $\Omega(n^{1.5}/\log n)$ time to deterministically simulate a 2-tape TM by a 1-tape TM with one-way input off-line (i.e., input ends with endmarkers).*

Before we proceed to prove the theorem, we need to prove a useful lemma. Consider a 1-tape TM M. Call M's input tape head h_1 and work tape head h_2. Let x_i be a block of input of a 1-tape TM M, and R be a tape segment on its work tape. We say that M *maps* x_i *into* R if h_2 never leaves tape segment R while h_1 is reading x_i. We say M maps x_i *onto* R if h_2 traverses the *entire* tape segment R while h_1 reads x_i.

We prove an intuitively straightforward lemma for one-tape machines with one-way input. A *crossing sequence* associated with work tape inter-square boundary p is $CS = ID_1, ID_2, \ldots$ with $ID_i =$ (state, input symbol) at the *ith* crossing of p at time t_i. The lemma states that a tape segment bordered by short c.s.'s cannot receive a lot of information without losing some. Formally:

Lemma 7.3.6 (Jamming Lemma) *Let the input string start with $x\# = x_1 x_2 \ldots x_k\#$, with the x_i's blocks of equal length C. Let R be a segment of M's storage tape and let l be an integer such that M maps each block $x_{i_1}, \ldots, x_{i_l}$ (of the x_i's) into tape segment R. The contents of the storage tape of M, at time $t_\#$ when $h_1(t_\#) = |x\#|$ and $h_1(t_\# - 1) = |x|$, can be reconstructed by using only the blocks $x_{j_1} \ldots x_{j_{k-l}}$ which remain from $x_1 \ldots x_k$ after deleting blocks $x_{i_1}, \ldots, x_{i_l}$, the final contents of R, the two final c.s.'s on the left and right boundaries of R, a description of M and a description of this discussion.*

Remark. Roughly speaking, if the number of missing bits $\sum_{j=1}^{l} |x_{i_j}|$ is greater than the number of added description bits ($< 3(|R| + 2|c.s.|) + O(\log |R|)$) then the Jamming Lemma implies that either $x = x_1 \ldots x_k$ is compressible or some information about x has been lost.

Proof of the Jamming Lemma. Let the two positions at the left boundary and the right boundary of R be l_R and r_R, respectively. We now simulate M. Put the blocks x_j of $x_{j_1} \ldots x_{j_{k-l}}$ in their correct positions on the input tape (as indicated by the h_1 values in the c.s.'s). Run M with h_2 staying to the left of R. Whenever h_2 reaches point l_R, the left boundary of R, we interrupt M and check whether the current ID matches the next ID, say ID_i, in the c.s. at l_R. Subsequently, using ID_{i+1}, we skip the input up to and including $h_1(t_{i+1})$, adjust the state of M to $M(t_{i+1})$, and continue running M. After we have finished left of R, we do the same thing right of R. At the end we have determined the appropriate contents of M's tape, apart from the contents of R, at $t_\#$ (i.e., the time when h_1 reaches $\#$). Inscribing R with its final contents from the reconstruction description gives us M's storage tape contents at time $t_\#$. Notice that although there are many unknown x_i's, they are never polled since h_1 skips over them because h_2 never goes into R. □

Proof of Theorem. The witness language L is defined by:

$$L = \{x_1@x_2@ \cdots @x_k \# y_1 @ \cdots @ y_l \# (1^i, 1^j) : x_i = y_j\}. \qquad (7.1)$$

Obviously, L can be accepted in linear time by a 2-tape one-way machine. Assume, by way of contradiction, that a deterministic one-way 1-tape machine M accepts L in $T(n) \notin \Omega(n^{1.5}/\log n)$ time. We derive a contradiction by showing that some incompressible string must have a too short description.

Assume, without loss of generality, that M writes only 0's and 1's in its storage squares and that $|M| \in O(1)$ is the number of states of M. Fix a constant C and the word length n as large as needed to derive the desired contradictions below and such that the formulas in the sequel are meaningful.

First, choose an incompressible string $x \in \{0,1\}^*$ of length $|x| = n$ (i.e., $K(x) \geq n$). Let x consist of the concatenation of $k = \sqrt{n}$ substrings, $x_1, x_2, \ldots, x_k$, each substring $\sqrt{n}$ bits long. Let

$$x_1@x_2@ \cdots @x_k \#$$

be the initial input segment on M's input tape. Let time $t_\#$ be the step at which M reads $\#$. If there are more than $k/2$ of the x_i's each mapped *onto* a contiguous tape segment of size at least n/C^3, then M requires $\Omega(n^{1.5}/\log n)$ time, which is a contradiction. Therefore, there is a set X of $k/2$ x_i's such that each of them is mapped *into* some tape segment of $\leq n/C^3$ contiguous tape cells. In the remainder of the proof we restrict attention to the x_i's in this set X. Order the elements of X according to the order of the left boundaries of the tape segments *into* which they are mapped. Let x_c be the median.

The idea of the rest of the proof is as follows. Our first intuition tells us: M could only somehow "copy" x_i's on its work tape, and then put y_j's on the work tape. There must be a pair of x_i and y_j that are separated by $O(n)$ distance since these blocks must occupy $O(n)$ space. Then if we ask M to check whether $x_i = y_j$, it has to spend about $\Omega(n^{1.5}/\log n)$ time. We convert this intuition into two cases: In the first case we assume that many x_i's in X are mapped (jammed) into a small tape segment R; that is, when h_1 (the input tape head) is reading them, h_2 (the storage tape head) is always in this small tape segment R. Intuitively, some information must have been lost. We show that then, contrary to assumption, x can be compressed (by the Jamming Lemma). In the second case, we assume there is no such 'jammed' tape segment, and that the records of the x_i's in X are "spread evenly" over the storage tape. In that case, we will arrange the y_j's so that there exists a pair (x_i, y_j) such that $x_i = y_j$ and x_i and y_j are mapped into tape segments that are far apart; i.e., of distance $O(n)$. Then we complete M's input with final index $(1^i, 1^j)$ so as to force M to match x_i against y_j. Now as in Example 7.3.2, either M spends too much time or we can compress x again, yielding a second contradiction and therefore $T(n) \in \Omega(n^{1.5}/\log n)$.

Case 1 (jammed). Assume there are k/C blocks $x_i \in X$ and a fixed tape segment R of length n/C^2 on the work tape such that M maps all of these x_i's into R. Let X' be the set of such blocks.

We will construct a short program which prints x. Consider the two tape segments of length $|R|$ to the left and to the right of R on the storage tape. Call them R_l and R_r, respectively. Choose positions p_l in R_l and p_r in R_r with the shortest c.s.'s in their respective tape segments. Both c.s.'s must be shorter than $\sqrt{n}/(C^2 \log n)$, for if the shortest c.s. in either tape segment is at least $\sqrt{n}/C^2 \log n$) long then M uses $\Omega(n^{1.5}/\log n)$ time, a contradiction. Let tape segment R'_l (R'_r) be the portion of R_l (R_r) right (left) of p_l (p_r).

Now, using the description of

- this discussion (including the text of the program below) and simulator M in $O(1)$ bits,

- the values of n, k, C, and the positions of p_l, p_r in $O(\log n)$ bits,

- $\{x_1, \ldots, x_n\} - X'$, using at most $n - n/C + O(\sqrt{n} \log n)$ bits; $O(\sqrt{n} \log n)$ bits are used for indicating indices,

- the state of M and the position of h_2 at time $t_\#$ in $O(\log n)$ bits,

- the two c.s.'s at positions p_r and p_l at time $t_\#$ in at most $2\sqrt{n}(|M| + O(\log n))$ bits, and

- the contents at time $t_\#$ of tape segment $R'_l R R'_r$ in at most $3n/C^2 + O(\log n)$ bits,

we can construct a program to check if a string y equals x by running M as follows.

Check if $|y| = |x|$. By the Jamming Lemma (using the above information as related to M's processing of the initial input segment $x_1@\cdots@x_k\#$) reconstruct the contents of M's work tape at time $t_\#$, the time h_1 gets to the first $\#$ sign. Divide y into k equal pieces and form $y_1@\cdots@y_k$. Simulate M, started on the input suffix, for each i,

$$y_1@\cdots@y_k\#(1^i, 1^i)$$

from time $t_\#$ onwards. By definition of L we have that M accepts for all i if and only if $y = x$.

This description of x requires not more than

$$n - \frac{n}{C} + \frac{3n}{C^2} + O(\sqrt{n} \log n) + O(\log n) \leq \gamma n$$

bits, for some positive constant $\gamma < 1$ and large enough C and n. However, this contradicts the incompressibility of x ($K(x) \geq n$).

Case 2 (not jammed). Assume that for each fixed tape segment R, with $|R| = n/C^2$, there are at most k/C blocks $x_i \in X$ mapped into R.

Fix a tape segment of length n/C^2 into which median x_c is mapped. Call this segment R_c. So at most k/C strings x_i in set X are mapped into R_c. Therefore, for large enough C ($C > 3$), at least $k/6$ of the x_i's in X are mapped into the tape right of R_c. Let the set of those x_i's be $X_r = \{x_{i_1}, \ldots, x_{i_{k/6}}\} \subset X$. Similarly, let $X_l = \{x_{j_1}, \ldots, x_{j_{k/6}}\} \subset X$, consist of $k/6$ strings x_i which are mapped into the tape left of R_c. Without loss of generality, assume $i_1 < i_2 < \cdots < i_{k/6}$, and $j_1 < j_2 < \cdots < j_{k/6}$.

Set $y_1 = x_{i_1}$, $y_2 = x_{j_1}$, $y_3 = x_{i_2}$, $y_4 = x_{j_2}$, and so forth. In general, for all integers m, $1 \leq m \leq k/6$,

$$y_{2m} = x_{j_m} \text{ and } y_{2m-1} = x_{i_m}, \tag{7.2}$$

We can now define an input prefix for M to be:

$$x_1@\cdots@x_k\#y_1@\cdots@y_{k/3}\#. \tag{7.3}$$

Claim 7.3.7 *There exist a pair $y_{2i-1}@y_{2i}$ which is mapped to a segment of size less than $n/(4C^2)$.*

Proof of Claim. If the claim is false then M uses $\Omega(n^{1.5}/\log n)$ time, a contradiction. □

Now this pair y_{2i-1}, y_{2i} is mapped to a segment with distance at least n/C^3 either to x_{i_m} or to x_{j_m}. Without loss of generality, let y_{2m-1}, y_{2m} be mapped n/C^3 away from x_{i_m}. So y_{2m-1} and x_{i_m} are separated by a region R of size n/C^3. Attach the input with suffix $(1^{i_m}, 1^{2m-1})$ to get the complete input to M

$$x_1@\cdots@x_k\#y_1@\cdots@y_{k/3}\#(1^{i_m}, 1^{2m-1}). \tag{7.4}$$

So at the time when M reads the second $\#$ sign, x_{i_m} is mapped into the left side of R and $y_{j_{2m-1}}$, which is equal to x_{i_m}, is mapped into the right side of R.

Determine a position p in R which has the shortest c.s. of M's computation on the input (7.4). If this c.s. is longer than $\sqrt{n}/(C^2 \log n)$ then M uses time $\Omega(n^{1.5}/\log n)$, a contradiction. Therefore, assume it has length at most $\sqrt{n}/(C^2 \log n)$. Then again we can construct a short program P, to accept *only* x by a 'cut and paste' argument, and show that it yields too short a description of x.

Using the description of

- this discussion (including the text of the program P below) and simulator M in $O(1)$ bits,

- the values of n, k, $C = n/k$, and the position of p in $O(\log n)$ bits,

- $n - \sqrt{n}$ bits for $X - \{x_{i_m}\}$,

- $O(\log n)$ bits for the index i_m of x_{i_m} to place it correctly on the input tape, together with

- $\leq \sqrt{n}/C$ bits for the c.s. of length $\sqrt{n}/(C^2 \log n)$ at p (assuming $C > |M|$),

we can construct a program to check if a string z equals x by running M as follows.

For a candidate input string z, program P first partitions z into $z_1 @ \cdots @ z_k$ and compares the appropriate literal substrings with the literally given strings in $\{x_1, \ldots, x_k\} - \{x_{i_m}\}$. The string x_{i_m} is given in terms of the operation of M; to compare the appropriate z_{i_m} with the x_{i_m}, we simulate M. First prepare an input according to form (7.3) as follows. Put the elements of $\{x_1, \ldots, x_k\} - \{x_{i_m}\}$ literally into their correct places on the input tape, filling the places for x_{i_m} arbitrarily. For the y_i's in (7.3) substitute the appropriate substrings z_i of candidate z according to scheme (7.2); i.e., use z_{j_m} for y_{2m} and $z_{i'_m}$ for y_{2m-1} ($1 \leq m \leq k/6$). Note that among these are all those substrings of candidate z which have not yet been checked against the corresponding substrings of x. Adding string $(1^{i_m}, 1^{2m-1})$ completes the input to M.

Using the c.s. at point p we run M such that h_2 always stays right of p (y_{2m-1}'s side). Whenever h_2 encounters p, we check if the current ID matches the corresponding one in the c.s.. If it does then we use the next ID of the c.s. to continue. If in the course of this simulation process M rejects or there is a mismatch (that is, when h_2 gets to p, M is not in the same state or h_1's position is not as indicated in the c.s.), then $z \neq x$. Note, that it is possible for M to accept (or reject) on the left of p (x_{i_m}'s side). However, once h_2 crosses p right-to-left for the last time M does not read z_{2m-1} substituted for y_{2m-1} any more and all other z_i's in prefix (7.3) are 'good' ones (we have already checked them). Therefore, if the crossing sequence of ID's at p of M's computation for candidate z match those of the prescribed c.s. then we know that M accepts. By construction the outlined program P accepts the string $z = x$. Suppose P also accepts $z' \neq x$. Then the described computation of M accepts for candidate z'. We can cut and paste the two computations of M with candidate strings z and z' using the computation with z left of p and the computation with z' right of p. Then string (7.4) composed from x and z' according to (7.2) is accepted by M. Since z and z' differ in block x_{i_m} this string is not in L as defined in (7.1): contradiction.

The description of x requires no more than

$$n - \sqrt{n} + \frac{\sqrt{n}}{C} + O(\log n) \leq n - \gamma\sqrt{n}$$

bits for some positive $\gamma > 0$ and large enough C and n. This contradicts the incompressibility of x $(K(x) \geq n)$ again.

Case 1 and Case 2 complete the proof that $T(n) \in \Omega(n^{1.5}/\log n)$. □

7.3.3   LOWER BOUNDS: MORE TAPES VERSUS FEWER TAPES

Although Barzdin [Ba68] and Paul [Pau79] pioneered the use of Kolmogorov complexity to prove lower bounds, the most influential paper is probably the one by Paul, Seiferas and Simon [PSS81], which was presented at the 1980 STOC. This was partly because [Pau79] was not widely circulated and, apparently, the paper by Barzdin [Ba68] did not even reach this community. The major goal of [PSS81] was "to promote the approach" of applying Kolmogorov complexity to obtain lower bounds. In [PSS81], apart from other results, the authors, with the aid of Kolmogorov complexity, remarkably simplified the proof of a well-known theorem of Aanderaa [Aan74]: *real-time* simulation of k tapes by $k - 1$ tapes is impossible for deterministic Turing machines. See also [PSNS88] for more improvements and simplifications.

In this model the Turing machine has k (work) tapes, apart from a separate input tape and (possibly) a separate output tape. This makes the machine for each $k \geq 1$ far more powerful than the model of Example 7.3.2, where the single tape is both input tape and work tape. For example, a 1-(work)tape Turing machine can recognize the palindromes of Example 7.3.2 in real-time $T(n) = n$ in contrast with $T(n) = \Omega(n^2)$ required in Example 7.3.2.

In 1982 Paul [Pau82], using Kolmogorov complexity, extended the results in [PSS81] to on-line simulation of real-time $k + 1$-tape Turing machines by k-tape Turing machines requires $\Omega(n(\log n)^{1/(k+1)})$ time. Duris, Galil, Paul, Reischuk [DGPR84] then improved the lower bound for the one versus two tape case to $\Omega(n \log n)$.

To simulate k tapes with 1 tape, the known (and trivial) upper bound on the simulation time was $O(n^2)$. Above lower bound decreased the gap with this upper bound only slightly. But in later developments w.r.t. this problem Kolmogorov complexity has been very successful. The second author, not using Kolmogorov complexity, reported in [Vit84a] a $\Omega(n^{1.5})$ lower bound on the time to simulate a single pushdown store on-line by one *oblivious* tape unit (that is, if the storage head movements are the same for all inputs of length n). However, using Kolmogorov complexity the technique worked also without the oblivious restriction, and yielded in quick succession [Vit85b,Vit85a], and the optimal results cited hereafter. Around 1983/1984, independently and in chronological order, Wolfgang Maass at UC Berkeley, the first author at Cornell and the second author at CWI Amsterdam, obtained a square lower bound on the time to simulate two tapes by one tape (deterministically), and thereby closed the gap between 1 tape versus k (w.l.o.g. 2) tapes. These lower bounds, and the following ones, were proven with the simulator an *off-line* machine with *one-way* input. All three relied on Kolmogorov complexity, and actually proved more in various ways. [3] Thus, Maass also obtained a nearly optimal result for nondeterministic simulation, [Maa85] exhibits a language that can be accepted by two deterministic one-head tape units in real-time, but for which a one-head tape unit requires $\Omega(n^2)$ time in the deterministic case, and $\Omega(n^2/(\log n)^2 \log\log n)$ time in the nondeterministic case. This lower bound was later improved by [LV88b] to $\Omega(n^2/\log n \log\log n)$ time using Maass' language, and by Galil, Kannan, and Szemeredi [GKS86] to $\Omega(n^2/\log^{(k)} n)$ (for any k, with $\log^{(k)}$ is the k-fold iterated logarithm) by an ingenious construction of a language whose computation graph does not have small separators. This almost closed the gap in the nondeterministic case. In their final combined paper, Li and Vitányi [LV88b] presented the following lower bounds, all by Kolmogorov complexity. To simulate 2 pushdown stores, or only 1 queue, by 1 deterministic tape requires $\Omega(n^2)$ time. Both bounds are tight. (Note that the 2 pushdown store result implies the 2 tape result. However, the 1 queue result is incomparable with either of them.) Further, 1-tape nondeterministic simulation of two pushdown stores requires $\Omega(n^{1.5}/\sqrt{\log n})$ time. This is almost tight because of [Li85b]. Finally, 1-tape nondeterministic simulation of one queue requires $\Omega(n^{4/3}/\log^{2/3} n)$ time. The corresponding upper bound of the last two simulations is $O(n^{1.5}\sqrt{\log n})$ in [Li85b]. In a successor paper, together with Luc Longpré, we have extended the above work with a comprehensive study stressing queues in comparison to stacks and tapes [LLV86]. There it was shown that a queue and a tape are not comparable; i.e., neither can simulate the other in linear time. Namely, simulating 1 pushdown store (and hence 1 tape) by 1 queue requires $\Omega(n^{4/3}/\log n)$, in both the deter-

---

[3] *Historical note.* A claim for an $\Omega(n^{2-\epsilon})$ lower bound for simulation of two tapes by both one deterministic tape and one nondeterministic tape was first circulated by W. Maass in August 1983, but did not reach Li and Vitányi. Maass submitted his extended abstract containing this result to STOC by November 1983, and this did not reach the others either. The final STOC paper of May 1984 (submitted February 1984) contained the optimal $\Omega(n^2)$ lower bound for the deterministic simulation of two tapes by one tape. In M. Li: 'On 1 tape versus 2 stacks,' Tech. Rept. TR-84-591, Dept. Comp. Sci., Cornell University, January 1984, the $\Omega(n^2)$ lower bound was obtained for the simulation of two pushdown stores by one deterministic tape. In: P.M.B. Vitányi, 'One queue or two pushdown stores take square time on a one-head tape unit,' Tech. Rept. CS-R8406, Centre for Mathematics and Computer Science, Amsterdam, March 1984, the $\Omega(n^2)$ lower bound was obtained for the simulation of two pushdown stores (or the simulation of *one* queue) by one deterministic tape. Maass's and Li's result were for off-line computation with one-way input, while Vitányi's result was for on-line computation. Li and Vitányi combined these and other results in [LV88b], while Maass published in [Maa85].

ministic and nondeterministic cases. Simulation of 1 queue by 1 tape was resolved above, and simulation of 1 queue by 1 pushdown store is trivially impossible. Nondeterministic simulation of 2 queues (or 2 tapes) by 1 queue requires $\Omega(n^2/(\log^2 n \log\log n))$ time, and deterministic simulation of 2 queues (or 2 tapes) by 1 queue requires quadratic time. All these results would be formidable without Kolmogorov complexity. In the previous section we have presented a proof of a weaker result of this flavor (Theorem 7.3.5).

A next step is to attack the similar problem with a *two-way input* tape. Maass and Schnitger [MS86] proved that when the input tape is two-way, two work tapes are better than 1 for *computing a function* (in contrast to recognizing a language). The model is a Turing machine with no output tape; the function value is written on the work tape(s) when the machine halts. It is interesting to note that they considered a matrix transposition problem, as considered in Paul's original paper. Apparently, in order to transpose a matrix, a lot of information needs to be shifted around which is hard for a single tape. [MS86] showed that transposing a matrix (with element size $O(\log n)$) requires $\Omega(n^{3/2}(\log n)^{-1/2})$ time on a 1-tape off-line Turing machine with an extra two-way read-only input tape. The first version of this paper (single authored by Maass) does not actually depend on Kolmogorov complexity, but has a cumbersome proof. The final Kolmogorov complexity proof was much easier and clearer. (This lower bound is also optimal [MS86].) This gives the desired separation of two tape versus one, because, with two work tapes, one can sort in $O(n \log n)$ time and hence do matrix transposition in $O(n \log n)$ time. Recently, Maass, Schnitger, and Szemeredi [MSS87] in 1987 finally resolved the question of whether 2 tapes are better than 1 with two-way input tape, for *language recognition*, with an ingenious proof. The separation language they used is again related to matrix transposition except that the matrices are Boolean and sparse (only $\log^{-2} n$ portion of nonzeros): $\{A^{**}B : A = B^t$ and $a_{ij} \neq 0$ only when $i, j = 0 \bmod(\log m)$ where m is the size of matrices $\}$. The proof techniques used combinatorial arguments rather than Kolmogorov complexity. There is still a wide open gap between the $\Omega(n \log n)$ lower bound of [MSS87] and the $O(n^2)$ upper bound. In [MS86] it was observed that if the Turing machine has a one-way output tape on which the transposed matrix can be written, transposition of Boolean matrices takes only $O(n^{5/4})$. Namely, with only one work tape and no output tape, once some bits have been written they can be moved later only by time wasting sweeps of the work tape head. In contrast, with an output tape, as long as the output data are computed in the correct order they can be output and don't have to be moved again. Using Kolmogorov complexity, in [Die87] Dietzfelbinger shows that transposition of Boolean matrices by Turing machines with two-way input tape, one work tape, and a one-way output tape requires $\Omega(n^{5/4})$ time, thus matching the upper bound for matrix transposition.

It turns out that the similar lower bound results for higher dimensional tapes are also tractable, and sometimes easier to obtain. The original paper [PSS81] contains such lower bounds. M. Loui proved the following results by Kolmogorov complexity. A *tree work tape* is a complete infinite rooted binary tree as storage medium (instead of a two-way infinite linear tape). A work tape head starts at the origin (the root) and can in each step move to the direct ancestor of the currently scanned node (if it is not the root) or to either one of the direct descendants. A *multihead tree machine* is a Turing machine with a tree work tape with $k \geq 1$ tree work tape heads. We assume that the finite control knows whether two work tape heads are on the same node or not. A *d-dimensional work tape* consists of nodes corresponding to d-tuples of integers, and a work tape head can in each step move from its current node to a node with each coordinate ± 1 of the current coordinates. Each work tape head starts at the origin which is the d-tuple with all zeros. A *multihead d-dimensional machine* is a Turing machine with a d-dimensional work tape with $k \geq 1$ work tape heads. M. Loui [Lou83] has shown that a multihead d-dimensional machine simulating a multihead tree machine on-line (both machines having a one-way input tape and one-way output tape) requires time $\Omega(n^{1+1/d}/\log n)$ in the worst case (this is optimal), and he proved the same lower bound as above for the case where multihead d-dimensional machine is made more powerful by allowing the work tape heads also to move from their current node to the current node of any other work tape head in a single step.

7.3.4   LOWER BOUNDS: MORE HEADS VERSUS FEWER HEADS

Again applying Kolmogorov complexity, Paul [Pau84] showed that two-dimensional two tape (with one head on each tape) Turing machines cannot on-line simulate two-dimensional Turing machines with two heads on one tape in real time. He was not able to resolve this problem for one-dimensional tapes, and, despite quite some effort, the following problem is open and believed to be difficult: Are two (one-dimensional) tapes, each with one head, as good as two heads on one (one-dimensional) tape? The following result, proved using Kolmogorov complexity, is intended to be helpful in separating these classes. A Turing machine with two one-head storage tapes cannot simulate a queue in both real time and with at least one storage head always within $o(n)$ squares from the start square [Vit84b]. (Thus, most prefixes of the stored string need to be shifted all the time, while storing larger and larger strings in the simulator, because the simulator must always be ready to reproduce the stored string in real-time. It would seem that this costs too much time, but this has not been proved yet.) To eventually exploit this observation to obtain the desired separation, J. Seiferas [Sei85] proved the following 'equal information distribution' property. For no c (no matter how large) is there a function $f(n) = o(n)$, such that every sufficiently long string x has a description y with the properties: $|y| = c|x|$, and if x' is a prefix of x and y' is any subword of y with $|y'| = c|x'|$ then $K(x'|y') < f(K(x))$.

Multihead finite automata and pushdown automata were studied in parallel with the field of computational complexity in the years of 1960's and 1970's. One of the major problems on the interface of the theory of automata and complexity is to determine whether additional computational resources (heads, stacks, tapes, etc.) increase the computational power of the investigated machine. In the case of multihead machines it is natural to ask whether $k + 1$ heads are better than k. A k-head finite (pushdown) automaton is just like a finite (pushdown) automaton except having k one-way heads on the input tape. Two rather basic questions were left open from the automata and formal language theory of 1960's:

(1) Rosenberg Conjecture (1965): $(k+1)$-head finite automata are better than k-head finite automata [Ros65,Ros66].

(2) Harrison-Ibarra Conjecture (1968): $(k+1)$-head pushdown automata are better than k-head pushdown automata. Or, there are languages accepted by $(k + 1)$-DPDA but not k-PDA [HI68].

In 1965, Rosenberg [Ros66] claimed a solution to problem (1), but Floyd [Flo68] pointed out that Rosenberg's informal proof was incomplete. In 1971 Sudborough [Sud74,Sud76], and later Ibarra and Kim [IK75] obtained a partial solution to problem (1) for the case of two heads versus three heads, with difficult proofs. In the 1976 Yao and Rivest [YR78] finally presented a full solution to problem (1). A different proof was also obtained by Nelson [Nel76]. Recently it was noted by several people, including Joel Seiferas and the authors, that the Yao-Rivest proof can be done very naturally and easily by Kolmogorov complexity: Let

$$L_b = \{w_1\# \cdots \#w_b\$w_b\# \cdots \#w_1 : w_i \in \{0, 1\}^*\}.$$

as defined by Rosenberg and Yao-Rivest. Let $b = \binom{k}{2} + 1$. With some effort one can see that L_b can be accepted by a $(k + 1)$-DFA. Assume that a k-FA M also accepts L_b. Let W be a long enough Kolmogorov random string and W be equally partitioned into $w_1w_2 \ldots w_b$. We say that the two w_i's in L_b are matched if there is a time such that two heads of M are in the two w_i's concurrently. Hence there is an i such that w_i is not matched. Then apparently, this w_i can be generated from $W - w_i$ and the positions of heads and states for M when a head comes in/out w_i, $K(w_i|W - w_i) = O(k \log n) < |w_i|/2$, contradiction.

The HI-conjecture, however, was open until the time of Applied Kolmogorov complexity. Several authors tried to generalize the Yao-Rivest method [Miy82,Miy83] or the Ibarra-Kim method [Chr86] to the k-PDA case, but only partial results were obtained. For the complete Odyssey of these efforts see the survey in [CL86]. With the help of Kolmogorov complexity, [CL86] presented a complete solution to the Harrison-Ibarra conjecture for the general case. The proof was constructive and quite simple compared to the partial solutions. The basic idea, ignoring the technical details, was generalized from the above proof we gave for the Rosenberg conjecture.

A related problem of whether a k-DFA can do string-matching was raised by Galil and Seiferas [GS81]. They proved that a six-head two-way DFA can do string-matching; i.e., accept $L = \{x\#y : x$ is a substring of $y\}$. In 1982, when the first author and Yaacov Yesha, then at Cornell, tried to solve the problem, we achieved a difficult and tediously long proof (many pages), by counting, that 2-DFA cannot do string matching. Later J. Seiferas suggested the use of Kolmogorov complexity, which shortened the proof to less than a page [LY86b]! By Kolmogorov complexity, proofs that 3-DFA cannot do string matching were also obtained [GL,Li85a].

7.3.5   LOWER BOUNDS: PARALLEL COMPUTATION AND BRANCHING-PROGRAMS

In this section, we demonstrate that the Kolmogorov complexity does not only apply to lower bounds in restricted Turing machines, it also applies to lower bounds in other general models, like parallel computing models.

Fast addition or multiplication of n numbers in parallel is obviously important. In 1985 Meyer auf der Heide and Wigderson [HW85] proved, using Ramsey theorems, that on priority PRAM, where the concurrent read and write are allowed, ADDITION (and MULTIPLICATION) requires $\Omega(\log n)$ parallel steps. Independently, a similar lower bound on addition was obtained by Israeli and Moran [IM] and Parberry [Par84]. All these lower bounds depend on inputs from infinite (or exponentially large) domains. However, in practice, we are often interested in small inputs of size $n^{O(1)}$ bits. It is known that addition of n numbers of $n^{1/\log\log n}$ bits each can be done in $O(\log n/\log\log n)$ time with $n^{O(1)}$ processors which is less than the $\Omega(\log n)$ time lower bound of [HW85]. In [LY86a] we applied Kolmogorov-complexity to obtain parallel lower bounds (and tradeoffs) for a large class of functions with arguments in small domains (including Addition, Multiplication ...) on priority PRAM. As a corollary, for example, we show that for numbers of $n^{O(1)}$ bits, it takes $\Omega(\log n)$ parallel steps to add them. This improved the results of [HW85,IM,Par84]. We next present this rather simple proof. Independently, Paul Beame obtained similar results, but using a different partition method. A proof of the above result was given in Example 7.3.2.

Formally, a priority PRAM consists of processors $P(i)$ $i = 1, 2, \ldots, n^{O(1)}$, and an infinite number of shared memory cells $C(i)$, $i = 1, 2, \ldots$. Each step of the computation consists of three parallel phases as follows. Each processor: (1) reads from a shared memory cell, (2) performs a computation, and (3) may attempt writing into some shared memory cell. In case of *write conflicts*, the processor with the minimum index succeeds in writing.

Theorem 7.3.8 *Adding n integers, each of $n^{O(1)}$ bits, requires $\Omega(\log n)$ parallel steps on a priority PRAM.*

Proof [LY86a]. Suppose that a priority PRAM M with $n^{O(1)}$ processors adds n integers in $o(\log n)$ parallel steps for infinitely many n's. The programs (maybe infinite) of M can be encoded into an oracle A. The oracle, when queried about (i, l), returns the initial section of length l of the program for $P(i)$. Fix a string $X \in \{0,1\}^{n^3}$ such that $K^A(X) \geq |X|$. Divide X equally into n parts $x_1, x_2, \ldots, x_n$. Then consider the (*fixed*) computation of M on input $(x_1, \ldots, x_n)$. We inductively define (with respect to X) a processor to be *alive* at step t in this computation if

(1) it writes the output; or

(2) it succeeds in writing something at some step $t' \geq t$ which is read at some step $t'' \geq t'$ by a processor who is alive at step t''.

An input is *useful* if it is read at some step t by a processor alive at step t. By simple induction on the step number we have: for a T step computation, the number of useful inputs and the number of processors ever alive are both $O(2^T)$.

It is not difficult to see that, given all the useful inputs and the set $ALIVE = \{(P(i), t_i) : P(i) \text{ was alive until step } t_i > 0\}$, we can simulate M to uniquely reconstruct the output $\sum_{i=1}^n x_i$. Since $T = o(\log n)$, we know $2^T = o(n)$. Hence there is an input x_{i_0} which is not useful. We need $O(2^T \log n^{O(1)}) = o(n \log n)$ bits to represent $ALIVE$. To represent $\{x_i : i \neq i_0\}$ we need $n^3 - n^2 + \log n$ bits, where $\log n$ bits are needed to indicate the index i_0 of the missing input. The total number of bits needed in the simulation is less than

$$J = n^3 - n^2 + O(n \log n) + O(\log n) < n^3.$$

But from these J bits we can find $\sum_{i=1}^n x_i$ by simulating M using the oracle A, and then reconstruct x_{i_0} from $\sum_{i=1}^n x_i$ and $\{x_i : i \neq i_0\}$. But then

$$K^A(X) \leq J < n^3.$$

This contradicts the randomness of X. □

Sorting is one of the most studied problems in computer science, due to its great practical importance. (It was also studied by Paul in [Pau79].) In 1979 Borodin, Fischer, Kirkpatrick, Lynch, and Tompa proved a time-space trade-off for comparison based sorting algorithms [BFK*79]. This was improved and generalized to a very wide class of sequential sorting algorithms by Borodin and Cook [BC80] defined as 'branching programs.' The proof involved difficult counting. In [RS82] Reisch and Schnitger used

Kolmogorov complexity, in one of their three applications, to simplify the well-known $\Omega(n^2/\log n)$ bound of Borodin and Cook [BC80] for the time-space trade-off in sorting with branching-programs. They also improved the lower bound in [BC80] to $\Omega(n^2 \log \log n/\log n)$.

7.3.6   LOWER BOUNDS: TIME-PROGRAM SIZE TRADE-OFF FOR SEARCHING A TABLE

"Is x in the table?" Let the table contain n keys. You can sort the table and then use binary search on the table; then, your program can be as short as $\log n$ bits, but you make about $\log n$ time (probes). Or you can do hashing; you can use a perfect hashing function $h(x) = \lfloor \frac{A}{Bx+C} \rfloor$ [CW79,Meh82], then your program may be as long as $O(n)$ bits since A, B and C must depend on the key space and they need to have $O(n)$ bits to make $h(x)$ perfect, but the search time is $O(1)$ probes. What is the size of the program? It is nothing but the Kolmogorov complexity.

Searching a table is one of the most fundamental issues in computer science. In a beautiful paper [Mai83] Mairson literally studied *the program complexity* of table searching procedures in terms of the number of bits that is required to write down such programs. In particular, he proved that a perfect hashing function of n keys needs $\Theta(n)$ bits to implement. He also provided the trade-offs between the time needed to search a table and the size of the searching program.

7.3.7   LOWER BOUNDS: VERY LARGE SCALE INTEGRATION

It should not be surprising that Kolmogorov complexity can be applied to VLSI lower bounds. Many VLSI lower bounds were based on the crossing sequence type arguments similar to that of Turing machines [LS81]. These sorts of arguments can be readily converted to much more natural and easier Kolmogorov complexity arguments like the one used in Example 7.3.2.

We use the model of Lipton and Sedgewick [LS81], which is a generalization of Thompson's Model [Tho79]. All lower bounds proved here also apply to the Thompson model. Roughly speaking, there are three main components in the model: (a) the (n-input, 1 output) Boolean function $f(x_1, x_2, \ldots, x_n)$ which is to be computed; (b) a synchronous circuit C, that computes f, which contains *and, or, not* gates of arbitrary fan-in and fan-out and with n fixed input gates (i.e., what is called where-oblivious) that are not necessarily on the boundary of the layout of C (the time an input arrives may depend on the data value); (c) and a VLSI (for convenience, rectangle) layout V that realizes C, where wires are of unit width and processors occupy unit squares. A central problem facing the VLSI designers is to find a C that computes a given f in time T and a VLSI layout of C with area A, minimizing say AT^2 as introduced by Thompson

[Tho79] and later generalized by [LS81].

The usual method used to prove $AT^2 = \Omega(n^2)$ lower bounds was roughly as follows. Draw a line to divide the layout into two parts, with about half the inputs on each part. Suppose the line necessarily cuts through ω wires, then $A > \Omega(\omega^2)$. Further, since for each time unit only one bit of information can flow through a wire, $T > I/\omega$ where I is the *amount* of information that has to be passed between the 2 parts. Then for each specific problem one only needs to show that $I = \Omega(n)$ for any division. Lipton and Sedgewick defined *crossing sequence* to be, roughly, the sequence of T tuples $(v_1, \ldots, v_\omega)$ where the ith tuple contains the values that appeared at the cut of width ω at step i.

Now it is trivial to apply our Kolmogorov complexity to simplify the proofs of *all* VLSI lower bounds obtained this way. Instead of complicated and non-intuitive counting arguments which involves *all* inputs, we now demonstrate how easily one can use one single Kolmogorov random string instead. The lower bounds before the work of [LS81] were for n-input and n-output functions, the Kolmogorov complexity can be even more trivially applied there. We only look at the harder n-input one-output problems stated in [LS81]. A sample question in [LS81]:

Example 7.3.9 (Pattern Matching) Given a binary text string of $(1 - \alpha)n$ bits and a pattern of αn bits, with $\alpha < 1$, determine if the pattern occurs in the text.

Proof Sketch. Let C implement pattern matching with layout V. Consider any cut of V of width ω which divides inputs into 2 halves. Now it is trivial that $I = \Omega(n)$ since for a properly arranged Kolmogorov random text and pattern this much information must pass the cut. This finishes the proof of $AT^2 \geq \Omega(n^2)$.   □

All other problems, Selection/Equality testing, DCFL, Factor Verification, listed in [LS81] can all be done similarly, even under the nondeterministic, or randomized circuits as defined in [LS81].

Some general considerations on VLSI lower bounds using Kolmogorov complexity were given by R. Cuykendall [Cuy84]. L.A. Levin and Y.Y. Itkis, [IL89], study the VLSI computation model under different information transmission assumptions, using Kolmogorov complexity [Lev83]. We use the self-delimiting variant of Kolmogorov complexity. In their model, if the speed of information transmission is superlinear, namely $max(K(d) - \log f(d)) < \infty$ for $f(d)$ the time for a signal to traverse a wire of length d, then a chip can be simulated by a chip in which all long wires have been deleted (which results in a considerable savings in required area). Note that $f(d) = \Omega(d \log^2 d)$ suffices, for f to satisfy the requirement, but not $f(d) = O(d)$.

7.3.8   LOWER BOUNDS: RANDOMIZED ALGORITHMS

We have seen that Kolmogorov complexity can be naturally applied to non-deterministic Turing machines. It is almost certain that it is useful for analyzing randomized algorithms. Indeed this is the case. In their paper about three applications of Kolmogorov complexity [RS82] Reisch and Schnitger, using Kolmogorov complexity, analyzed the probabilistic routing algorithm in n-dimensional cubes of Valiant and Brebner [VB81].

In 1983 Paturi and Simon generalized the deterministic lower bounds previously proved by [Aan74,Pau84,Pau82,PSS81] etc., to probabilistic machines. This is based on the following elegant idea (based on a note of, and discussions with, R. Paturi). As we mentioned before, all the Kolmogorov complexity proofs depend on only a fixed Kolmogorov random string α. If the lower bound fails, then this incompressible string can be compressed, hence a contradiction. [PS83] proved a version of the Symmetry of Information Lemma we stated in a previous section. They show that for a sequence of random coin tossing, the probability that this sequence of random coin tossing bits, β, contains much information about α is vanishingly small. Observe that if α is Kolmogorov random relative to the coin tossing sequence β, then the old deterministic argument would just fall through with β as an extra useless input (or oracle). And note that many such α exists. Hence, (ignoring technical details) using this idea and careful construction of the input for the probabilistic simulator, it was shown that, on the average, the probabilistic simulator would not give any advantage in reducing the computation time.

Remark. Similar ideas were expressed earlier by Levin who called the general principle involved "Law of Information Conservation" [Lev74]. See for later developments also [Lev84].

7.3.9   LOWER BOUNDS: FORMAL LANGUAGE THEORY

The classic introduction to formal language theory is [HU79]. An important part of formal language theory is deriving a hierarchy of language families. The main division is the Chomsky hierarchy, with regular languages, context-free languages, context-sensitive languages and recursively enumerable languages. The common way to prove that certain languages are not regular [not context-free] is by using "pumping" lemma's; i.e., the uvw-lemma [$uvwxy$-lemma]. However, these lemma's are complicated to state and cumbersome to prove or use. In contrast, below we show how to replace such arguments by simple, intuitive and yet rigorous, Kolmogorov complexity arguments. We present some material from our paper [LV89b]. Without loss of generality, languages are infinite sets of strings over a finite alphabet.

Regular languages coincide with the languages accepted by finite automata (FA). Another way of stating this is by the Myhill-Nerode Theorem:

each regular language over alphabet V consists of the union of some equivalence classes of a right-invariant equivalence relation on $V^* (= \bigcup_{i \geq 0} V^i)$ of finite index. Let us give an example of how to use Kolmogorov complexity to prove non-regularity. We prove that $\{0^k 1^k : k \geq 1\}$ is not regular. To the contrary, suppose it is regular. Fix k with $K(k) \geq \log k$, with k large enough to derive the contradiction below. The state q of the accepting FA after processing 0^k is, up to a constant, a description of k. Namely, by running the FA, starting from state q, on a string consisting of 1's, it reaches its first accepting state precisely after k 1's. Hence, there is a constant c, depending only on the FA, such that $\log k < c$, which is a contradiction. We generalize this observation, actually a Kolmogorov-complexity interpretation of the Myhill-Nerode Theorem, as follows. (In lexicographic order short strings precede long strings.)

Lemma 7.3.10 (KC-Regularity) *Let L be regular. Then for some constant c depending only on L and for each string x, if y is the nth string in the lexicographical order in $L_x = \{y : xy \in L\}$ (or in the complement of L_x) then $K(y) \leq K(n) + c$.*

  Proof. Let L be a regular language. A string y such that $xy \in L$, for some x and n as in the Lemma, can be described by

(a) This discussion, and a description of the FA that accepts L,

(b) The state of the FA after processing x, and the number n.

□

The KC-regularity lemma can be applied whenever the pumping lemma can be applied. It turns out that the converse of our Lemma also holds. Together this gives a Kolmogorov complexity *characterization* of regular languages [LV89b]. Therefore, the above Lemma also applies to situations when the normal pumping lemma(s) do(es) not apply. Further it is easier and more intuitive than pumping lemmas. For example:

Example 7.3.11 We prove that $\{1^p : p \text{ is prime }\}$ is not regular. Consider the string xy consisting of p 1's, where p is the $(k+1)$th prime. Set in the lemma x equal $1^{p'}$ with p' the kth prime, so $y = 1^{p-p'}$, and $n = 1$. It follows that $K(p - p') = O(1)$. Since the differences between the consecutive primes rise unbounded, this implies that there is an unbounded number of integers of Kolmogorov complexity $O(1)$. Since there are only $O(1)$ descriptions of length $O(1)$, we have a contradiction.

Example 7.3.12 (Exercise $3.1(h^*)$ **in [HU79])** Prove that $L = \{xx^R w : x, w \in \{0,1\}^+\}$ is not regular. Fix x such that $K(x) \geq |x|$. Consider prefix $(01)^{3 \log |x|} x$. The first string with this prefix in L is

$(01)^{3\log|x|}xx^R(10)^{3\log|x|}0$. By the KC-regularity lemma,
$K(x^R(10)^{3\log|x|}0) \leq K(1) + c$, contradiction.

Example 7.3.13 (Exercise 3.6* **in [HU79])** Prove that
$L = \{0^i1^j : GCD(i,j) = 1\}$ is not regular. Obviously, L is regular if and
only if its complement is regular. For each prime p, the string 0^p1^p is the
second word in the complement with prefix 0^p. Hence by the Lemma there
is a constant c such that for all p we have $K(p) < c$, which is a contradiction.

Similar general lemmas can also be proved to separate DCFL's from
CFL's. Previous proofs that a CFL is not a DCFL often use *ad hoc* methods.
We refer the interested readers to [LV89b].

It is known that linear context-free languages can be recognized on-line
by a one work tape Turing machine in $O(n^2)$ time. This is due to Kasami.
Gallaire using a very complicated counting argument and *de Bruijn sequences* [Gal69] proved

Theorem 7.3.14 *A multitape turing machine requires* $\Omega(n^2/\log n)$ *time
to on-line recognize linear context-free languages.*

The following elegant and short proof is due to Seiferas [Sei86] who, using
Kolmogorov complexity, significantly simplified Gallaire's proof.
Proof [Seiferas]. Define a linear context-free language:

$$L = \{y\#x_1@x_2@\cdots@x_k : k \geq 0, \ x_1,\ldots,x_k, \ y \in \{0,1\}^*,$$
$$\exists i < k \ \exists y = ux_i^Rv, \ u,v \in \{0,1\}^*\}.$$

Assume M accepts L in $O(n^2/\log n)$ steps on-line. Choose y such that
$y = \lfloor n/2 \rfloor$ and $K(y) \geq |y|$. We specify a hard input of length n for M using
y.

The idea of the proof is to construct an input prefix

$$y\#x_1@\cdots@x_{i-1}@$$

such that no $x_l(1 \leq l \leq i)$ is the reverse of a substring of y and yet each x_l
is hard enough to make M run ϵn steps, where $\epsilon > 0$ does not depend on
n. Now if $i = O(n/\log n)$, M is forced to run $\Omega(n^2/\log n)$ steps. Our task
is to search for such x_l's. We first prove two easy lemmas.

Lemma 7.3.15 *If* $K(x) \geq |x|$ *then* x *has no repetition longer than* $3\log_2|x|$.

Lemma 7.3.15 is Fact 7.2.8 proved in Section 7.2.

Lemma 7.3.16 *If a string has no repetition of length* m, *then it is uniquely
determined by the set of its substrings of length* $m + 1$; *i.e., then it is the
unique string with no repetition of length* m *and with precisely its set of
substrings of length* $m + 1$.

Proof of Lemma 7.3.16. Let S be the set of substrings of x of length $m + 1$. Let $a, b \in \{0, 1\}$, and $u, v, w \in \{0, 1\}^*$. The prefix of x of length $m + 1$ corresponds uniquely to the $ua \in S$ such that for no b is bu in S. For any prefix vw of x with $|w| = m$, there is a unique b such that $wb \in S$. Hence, the unique prefix of length $|vw| + 1$ is vwb. Hence the lemma follows by induction.   □

We continue our proof of the theorem. Assume inductively that $x_1, \ldots, x_{i-1}$ have been chosen as required, so that the input prefix $y\#x_1@ \cdots @x_{i-1}@$ does not yet belong to L. By Lemmas 7.3.15 and 7.3.16, we let $m = 3\log_2 n$ so that y is uniquely determined by its set of substrings of length m. We claim that, for each $i = 1, 2, \ldots, n/m$, there is an x_i of length m that is not a reverse substring of y such that appending $x_i@$ to the input requires at least $t = \epsilon n$ many additional steps by M, where $\epsilon > 0$ does not depend on n. This proves the Theorem. Assume that the claim is not true for some i. Then we can devise a short description of all the length m substrings of y, and hence, by Lemma 7.3.16, of y. Simulate M with input $y\#x_1@ \cdots @x_{i-1}@$, and record the following information at the time t_0 when M reads the last @ sign

- this discussion,

- work tape contents within distance t of the tape heads,

- specification of M, length n, current state of M, location of M's heads.

With this information, one can easily search for all x_i's such that $|x_i| = m$ and $x_i{}^R$ is a substring of y as follows. Run M from time t_0, using the above information and with input suffix $x_i@$, for t steps. By assumption, if M accepts or uses more than t steps, then x_i is a reverse substring of y. If ϵ is small in terms of M, and if n is large, then all above information adds up to fewer than $\lfloor n/2 \rfloor$ bits, a contradiction.   □

7.3.10   LOWER BOUNDS: WHICH METHOD TO USE?

Instead of attempting to answer this difficult question, we present a problem with three proofs: one by counting, one by probabilistic argument, one by Kolmogorov complexity. The problem and first two proofs are taken from a beautiful book by Erdös and Spencer [ES74].

A *tournament* T is a complete directed graph. That is, for each pair of vertex u and v in T, exactly one of edges (u, v), (v, u) is in the graph. Given a tournament T of n nodes $\{1, \cdots, n\}$, fix any standard effective coding, denote by $c(T)$, using $n(n - 1)/2$ binary bits, one bit for each edge. The bit of edge (u, v) is set to 1 if and only if (u, v) is in T. The next theorem and the first two proofs are from the first example in [ES74].

Theorem 7.3.17 *If $v(n)$ is the largest integer such that every tournament on $\{1, \cdots, n\}$ contains a transitive subtournament on $v(n)$ players, then $v(n) \leq 1 + \lceil 2 \log_2 n \rceil$.*

Remark. This theorem was proved first by P. Erdös and L. Moser in 1964. P. Stearns showed by induction that $v(n) \geq 1 + \lceil \log_2 n \rceil$.

Proof by Counting. Let $v = 2 + \lceil 2 \log_2 n \rceil$. Let $\Gamma = \Gamma_n$ be the class of all tournaments on $\{1, \cdots, n\}$, $\Gamma' = $ the class of tournaments on $\{1, \cdots, n\}$ that do contain a transitive subtournament on v players. Then

$$\Gamma' = \bigcup_A \bigcup_\sigma \Gamma_{A,\sigma}$$

where $A \subseteq \{1, \cdots, n\}$, $|A| = v$, σ is a permutation on A, and $\Gamma_{A,\sigma}$ is the set of T such that $T|A$ is generated by σ. If $T \in \Gamma_{A,\sigma}$, the $\binom{v}{2}$ games of $T|A$ are determined. Thus

$$|\Gamma_{A,\sigma}| = 2^{\binom{n}{2} - \binom{v}{2}}$$

and by elementary estimates

$$|\Gamma'| \leq \sum_{A,\sigma} 2^{\binom{n}{2} - \binom{v}{2}} = \binom{n}{v} v! 2^{\binom{n}{2} - \binom{v}{2}} < 2^{\binom{n}{2}} = |\Gamma|.$$

Thus $\Gamma - \Gamma' \neq \emptyset$. That is, there exists $T \in \Gamma - \Gamma'$ not containing a transitive subtournament on v players.   □

Proof by Probabilistic Argument. Let $v = 2 + \lceil 2 \log_2 n \rceil$. Let $\Gamma = \Gamma_n$ be the class of all tournaments on $\{1, \cdots, n\}$. Also let $A \subseteq \{1, \cdots, n\}$, $|A| = v$, σ be a permutation on A. Let $\mathbf{T} = T_n$ be a random variable. Its values are the members of Γ where for each $T \in \Gamma$, $Pr(\mathbf{T} = T) = 2^{-\binom{n}{2}}$. That is, all members of Γ are equally probable values of $\mathbf{T}$. Then

$$Pr(\mathbf{T} \text{ contains a transitive subtournament on } v \text{ players})$$

$$\leq \sum_A \sum_\sigma Pr(\mathbf{T}|A \text{ generated by } \sigma)$$

$$= \binom{n}{v} v! 2^{-\binom{v}{2}} < 1.$$

Thus some value T of $\mathbf{T}$ does not contain a transitive subtournament on v players.   □

Proof by Kolmogorov Complexity. Fix $T \in \Gamma_n$ such that $K(c(T)|n, v(n)) \geq |c(T)| = n(n-1)/2$. Suppose $v(n) = 2 + \lceil 2 \log_2 n \rceil$ and let S be the transitive tournament of $v(n)$ nodes. We effectively re-code $c(T)$ as follows in less than $|c(T)|$ bits, and hence we obtain a contradiction, by:

- List in order of dominance the index of each node in S in front of $c(T)$, using $2(\lceil \log_2 n \rceil)^2 + 2\lceil \log_2 n \rceil + |c(T)|$ bits;

- Delete all bits from $c(T)$ for edges in between nodes in S to save $2(\lceil \log_2 n \rceil)^2 + 3\lceil \log_2 n \rceil + 1$ bits.

□

7.3.11 LOWER BOUNDS: OPEN QUESTIONS

We list a few open questions considered to be interesting and possibly solvable by Kolmogorov complexity.

(1) Can k-DFA do string-matching [GS81]?

(2) Are 2 heads on one (one-dimensional) tape better than two (one-dimensional) tapes each with one head?

(3) Prove tight, or $\Omega(n^{1+\epsilon})$, lower bound for simulating two tapes by one for off-line Turing machines with an extra two-way input tape.

7.4 Resource-Bounded Kolmogorov Complexity and The Structure of Complexity Classes

Here we treat several notions of resource-bounded Kolmogorov complexity, with applications ranging from the P = NP question to factoring integers and cryptography. Several authors early on suggested the possibility of restricting the power of the device used to compress strings. Says Kolmogorov [Kol65] in 1965:

> "The concept discussed ... does not allow for the "difficulty" of preparing a program p for passing from an object x to an object y. ... [some] object permitting a very simple program; i.e., with very small complexity $K(x)$ can be restored by short programs only as the result of computations of a thoroughly unreal nature. ... [this concerns] the relationship between the necessary complexity of a program and its permissible difficulty t. The complexity $K(x)$ that was obtained [before] is, in this

case, the minimum of $K^t(x)$ on the removal of the constraints on t."

The earliest use of resource-bounded Kolmogorov complexity we know of is Barzdin's 1968 result [Ba68] cited earlier. Time-limited Kolmogorov complexity was applied by Levin [Lev73] in relation to his independent work on NP-completeness, and further studied in [Lev84]. Adleman investigated such notions [Adl79], in relation to factoring large numbers. Resource-bounded Kolmogorov complexity was extensively investigated by Daley [Dal77], Ko [Ko83], and Hartmanis [Har83]. Sipser [Sip83] used time-limited Kolmogorov complexity to show that the class BPP (problems which can be solved in polynomial time with high probability) is contained in the polynomial time hierarchy: $\text{BPP} \subseteq \Sigma_4^P \cap \Pi_4^P$. (Gács improved this to $\text{BPP} \subseteq \Sigma_2^P \cap \Pi_2^P$.) We treat the approaches of Adleman, Bennett and Hartmanis in more detail below. Let us note here that there is some relation between the approaches to resource-bounded Kolmogorov complexity by Adleman [Adl79], Levin [Lev84], and Bennett [Ben88].

7.4.1 POTENTIAL

In an elegant paper [Adl79], Adleman formulates the notion of *potential* as the amount of time that needs to be pumped in a number by the computation that finds it. Namely, while constructing a large composite number from two primes we spend only a small amount of time. However, to find the primes back may be difficult and take lots of time. Is there a notion of storing potential in numbers with the result that high-potential primes have relatively low-potential products? Such products would be hard to factor, because all methods must take the time to pump the potential back. Defining the appropriate notion, Adleman shows that if factoring is not in P then this is the reason why.

Formally, for all integer $k \geq 0$, for all $x \in \{0,1\}^*$ (for all $y \in \{0,1\}^*$), *x is k-potent (with respect to y)* if and only if there is a program p of size $\leq k \log |x|$, which with blanks (y) as input halts with output x in less than or equal to $|x|^k$ steps. (Recall that $|x|$ is the length of x and x can mean the positive integer x or the xth binary string.)

Example 7.4.1 For almost all $n \in N$, 1^n is 2-potent. Namely, $|1^n| = n$ and $|n| \sim \log n$. Then it is not difficult to see that, for each large enough n, there is a program p, $|p| < 2 \log n$, that computes 1^n in less than n^2 steps.

Example 7.4.2 For all k, for almost all incompressible x, x is not k-potent. This follows straightaway from the definitions.

Example 7.4.3 Let u be incompressible. If $v = u + 1^{666}$, where "+" denotes "exclusive or", then v is incompressible, but also v is 1-potent with respect to u.

Lemma 7.4.4 (Adleman) *For $x \in \{0,1\}^n$ and $k \in N$, define $y = y_1 y_2 \cdots y_m$ with $m = 2^n$, for some n, such that, for $1 \leq i \leq m$, $y_i = \bar{i}$ if i is k-potent with respect to x and $y_i = \epsilon$ otherwise. Let x' be the self-delimiting version of x. Then, for all k, the function $f_k(x'1^n) = y$ is computable in polynomial time.*

Proof. There are at most $2 \cdot 2^{k|n|} \sim 2n^k$ programs of length $\leq k|n|$. By simulating all such programs (one after the other) on input x for at most n^k steps the result is obtained.   □

We informally state two results proved by Adleman.

Theorem 7.4.5 (Adleman) *Factoring is difficult if and only if multiplication infinitely often takes highly potent numbers and produces relatively low potent numbers.*

Theorem 7.4.6 (Adleman) *With respect to the $P = NP$ question: $SAT \in NP - P$ if and only if for all k there exist infinitely many $\phi \in SAT$ such that [for all T, if truth assignment T satisfies ϕ then T is not k-potent w.r.t. ϕ]*

7.4.2 LOGICAL DEPTH

C. Bennett has formulated an intriguing notion of logical depth [Ben88, Ben87]. Kolmogorov complexity helps to define individual information and individual randomness. It can also help to define a notion of "individual computational complexity" of a finite object. Some objects are the result of long development (=computation) and are extremely unlikely to arise by any probabilistic algorithm in a small number of steps. Logical depth is the necessary number of steps in the deductive or causal path connecting an object with its plausible origin. Concretely, the time required by a universal computer to compute an object from its maximally compressed description. Formally, (in P. Gács' reformulation, using the Solomonoff-Levin approach to *a priori* probability):

$$depth_\epsilon(x) = min(t : \mathbf{m}_t(x) \geq \epsilon \mathbf{m}(x))$$

Here $\mathbf{m}_t$ is the universal t-bounded probability, and $\mathbf{m}$ is the universal probability. Thus, the depth of a string x is at least t with confidence $1 - \epsilon$ if the conditional probability that x arises in t steps *provided it arises* is less than ϵ. (One can also formulate logical depth in terms of shortest programs and running times [Ben88], or Example below.) According to Bennett, quoted in [Cha77]: "A structure is deep, if it is superficially random but subtly redundant, in other words, if almost all its algorithmic probability is contributed by slow-running programs. ... A priori the most probable explanation of 'organized information' such as the sequence of bases in a naturally occurring DNA molecule is that it is the product of an extremely long biological process."

Example 7.4.7 (Bennett) Bennett's original definition: Fix, as usual, an optimal universal machine U. A string $x \in \{0,1\}^*$ is logical (d,b)-deep, or "d-deep at confidence level 2^{-b}", if every program to compute x in time $\leq d$ is compressible by at least b bits.

The notion is intended to formalize the idea of a string for which the null hypothesis that it originated by an effective process of fewer than d steps, is as implausible as tossing b consecutive heads. Depth should be stable; i.e., no trivial computation should be able to transform a shallow object into a deep one.

Theorem 7.4.8 (Bennett) *Deep strings cannot quickly be computed from shallow ones. More precisely, There is a polynomial $p(t)$ and a constant c, both depending on U, such that, if x is a program to compute y in time t, and if x is less than (d,b)-deep, then y is less than $(d+p(t), b+c)$-deep.*

Example 7.4.9 (Bennett) Similarly, depth is reasonably machine independent. If U, U' are two optimal universal machines, then there exists a polynomial $p(t)$ and a constant c, both depending on U, U', such that $(p(d), b+c)$-depth on either machine is a sufficient condition for (d,b)-depth on the other.

Example 7.4.10 (Bennett) The distinction between depth and information: consider the numbers k and Ω ($k = k_1 k_2 \ldots$ with $k_i = 1$ if the universal TM U halts on the *ith* program, and $k_i = 0$ otherwise, $\Omega = \sum_x 2^{-K(x)}$, $K(x)$ the self-delimiting Kolmogorov complexity of x) k and Ω encode the same information, viz. solution to the halting problem. But k is deep and Ω shallow. Because Ω encodes the halting problem with maximal density (the first 2^n bits of k can be computed from the first $n + O(\log n)$ bits of Ω) it is recursively indistinguishable from random noise and practically useless. The time required to compute an initial segment of k from an initial segment of Ω increases faster than any computable function. Namely, Barzdin' [Ba68] showed that the initial segments of k are compressible to the logarithm of their length if unlimited time is allowed for decoding, but can only be compressed by a constant factor if any recursive bound is imposed on the decoding time.

7.4.3   Generalized Kolmogorov Complexity

Below we partly follow [Har83]. Assume that we have fixed a universal Turing machine U with an input tape, work tapes and an output tape. To say a string x is computed from a string z (z is a *description* of x) means that U started with z on its input tape halts with x on its output tape.

Remark. In order to be accurate in the reformulations of notions in the Examples below, we shall assume w.l.o.g. that the set of programs for which U halts is an effective prefix code; no such program is the prefix of any other such program. That is, we use self-delimiting descriptions as described in a previous section.

In the following we distinguish the main parameters we have been able to think of: compression factor, time, space, and whether the computation is inflating or deflating. A string x has *resource bounded* Kolmogorov complexity $_U^{UP}(K, T, S)$ if x can be computed from a string z, $|z| \leq K \leq |x|$, in $\leq T$ steps by U using $\leq S$ space on its work tape. A string x of length n is in complexity class $K_U^{UP}[k(n), t(n), s(n)]$ if $K \leq k(n)$, $T \leq t(n)$ and $S \leq s(n)$. Thus, we consider a computation that *inflates* z to x. A string x has *resource bounded* Kolmogorov complexity $_U^{DOWN}(K, T, S)$ if some description z of x can be computed from x, $|z| \leq K \leq |x|$, in $\leq T$ steps by U using $\leq S$ space on its work tape. Here we consider a computation that *deflates* x to z. A string x of length n is in complexity class $K_U^{DOWN}[k(n), t(n), s(n)]$ if $K \leq k(n)$, $T \leq t(n)$ and $S \leq s(n)$. Clearly,

$$K_U^{DOWN}[k(n), \infty, \infty] = K_U^{UP}[k(n), \infty, \infty] = K[k(n)],$$

$k(n)$ fixed up to a constant, with $K[k(n)]$ is (with some abuse of notation) the class of binary strings x such that $K(x) \leq k(n)$. (Here we denote by K the self-delimiting Kolmogorov complexity).

It follows immediately by the Hennie-Stearns simulation of many work tapes by two work tapes, that there is a U with two work tapes such that, for any multitape universal Turing machine V, there is a constant c such that

$$K_V^{UP}[k(n), t(n), s(n)] \subseteq$$
$$K_U^{UP}[k(n) + c, c \cdot t(n) \log t(n) + c, cs(n) + c].$$

Thus, henceforth we drop the subscripts because the results we derive are invariant up to such small perturbations. It is not difficult to prove, however, that larger perturbations of the parameters separate classes. For instance,

$$K^{UP}[\log n, \infty, n^2] \subset K^{UP}[\log n, \infty, n^2 \log n]$$
$$K^{UP}[\log n, \infty, n^2] \subset K^{UP}[2 \log n, \infty, n^2].$$

The obvious relation between inflation and deflation is:

$$K^{UP}[k(n), t(n), \infty] \subseteq K^{DOWN}[k(n), t(n)2^{k(n)}, \infty],$$

$$K^{DOWN}[k(n), t(n), \infty] \subseteq K^{UP}[k(n), t(n)2^n, \infty],$$

(there are at most $2^{k(n)}$ [2^n] possibilities to try). Our $K^{UP}[f(n), g(n), h(n)]$ will also be written as $K[f(n), g(n), h(n)]$ to be consistent with the literature as in [Har83].

In his Ph.D. thesis [Lon86] Longpré analyzed the structure of the different generalized Kolmogorov complexity sets, with different time and space bounds (the UP version). Longpré builds the resource hierarchies for Kolmogorov complexity in the spirit of classical time and space complexity hierarchies. He related further structural properties to classical complexity. He also extended Martin-Löf's results to generalized Kolmogorov complexity: the space bounded Kolmogorov complexity random strings pass all statistical tests which use less space than the space bound. Finally, he shows how to use Kolmogorov randomness to build a pseudo-random number generator that passes Yao's test [Yao82].

Example 7.4.11 (Potency) Adleman's potency [Adl79], can now be reformulated as: $x \in \{0,1\}^*$, $|x| = n$, is k-potent if $x \in K[k \log n, n^k, \infty]$.

Example 7.4.12 (Hitting Time) Related to the notions of potential and logical depth is Levin's concept of *hitting time complexity* Kt [Lev84]. In this framework we formulate it by: $x \in \{0,1\}^*$, has Kt-complexity $Kt(x) = m$ if $x \in K[m - \log t, t, \infty]$, m minimal.

Example 7.4.13 (Hartmanis) The sparse set

$$SAT \cap K[\log n, n^2, \infty]$$

is a Cook-complete set for all other sparse sets in NP.

In [Har83] these and similar results are derived for PSPACE and sets of other densities. It is also used to give new interpretations to oracle constructions, and to simplify previous oracle constructions. This leads to conditions in terms of Kolmogorov complexity under which there exist NP complete sets that are not polynomial-time isomorphic, as formulated in [BH77]. In [HH86] a characterization of the P = NP question is given in terms of time-bounded Kolmogorov complexity and relativization. Earlier, Adleman with [Adl79] established a connection, namely, NP $\neq$ P exactly when NP machines can "manufacture" randomness. Following this approach, Hemachandra [Hem86] obtains unrelativized connections in the spirit of [HH86].

Example 7.4.14 (Hartmanis) Hartmanis noticed the following interesting fact: a polynomial machine cannot compute from simple input complicated strings and hence cannot ask complicated questions to an oracle A. Using this idea, he constructed several very elegant oracles. As an example, we construct the Baker-Gill-Solovay oracle A such that $P^A \neq NP^A$: By diagonalization, choose $C \subseteq \{1^{2^n} : n \geq 1\}$ and $C \in DTIME[n^{\log n}] - P$. For every n such that $1^{2^n} \in C$ put the first string of length 2^n from

$$K[\log n, n^{\log n}, \infty] - K[\log n, n^{\log \log n}, \infty]$$

in A. Clearly, $C \in \mathrm{NP}^A$. But C cannot be in P^A since in polynomial time, a P^A-machine cannot ask any question about any string in A. Hartmanis also constructed two others including a random sparse oracle A such that $\mathrm{NP}^A \neq \mathrm{P}^A$ with probability 1.

Example 7.4.15 (Longpré, Natarajan) [Lon86,Nat88]. It was noticed that Kolmogorov complexity can be used to obtain space complexity hierarchies in Turing machines. Also it can be used to prove certain immunity properties. For example, one can prove that if $\lim_{n \to \infty} \frac{S(n)}{S'(n)} = 0$, then for any universal machine U, if $S'(n) \geq n$ is a nondecreasing function, $f(n)$ is a function not bounded by any constant and computable in space $S(n)$ by u, we have the complement of $K_U[f(n), \infty, S'(n)]$ is DSPACE$[S(n)]$-immune, for large n.

7.4.4   GENERALIZED KOLMOGOROV COMPLEXITY APPLIED TO STRUCTURAL PROOFS

Generalized Kolmogorov complexity turns out to be an elegant tool for studying the structure in complexity classes. The first such applications are probably due to Hartmanis, as we discussed in previous section. Other work in this area includes [All87,BB86,Pet80]. In this section we try to present some highlights of the continuing research in this direction. We will present several excellent constructions, and describe some constructions in detail.

Example 7.4.16 (An Exponentially Low Set Not In P) A set A is exponentially low if $\mathrm{E}^A = \mathrm{E}$, where $\mathrm{E} = DTIME[2^{cn}]$. Book, Orponen, Russo, and Watanabe [BORW88] constructed an exponentially low set A which is not in P. We give this elegant construction in detail. Let $K = K[n/2, 2^{3n}, \infty]$. Let $A = \{x : x$ is the lexicographically least element of $\bar{K}$ of length $2^{2^{\cdot^{\cdot^2}}}$ (stack of m 2's), for some $m > 0\}$. Obviously $A \in \mathrm{E}$. Further A is not in P since otherwise we let $A = L(M)$ and for $|x| \gg |M|$ and $x \in A$ we would have that $x \in K$, a contradiction. We also need to show that $\mathrm{E}^A = \mathrm{E}$. To simulate a computation of E^A by an E machine: if a query to A is of correct length (stack of 2's) and shorter than cn for a small constant c, then just do exhaustive search to decide. Otherwise, the answer is "no" since (1) a string of wrong length (no stack of 2's) is not in A and (2) a string of length greater than cn is not even in $\bar{K}$. (2) is true since the query string can be calculated from the input of length n and the exponentially shorter previous queries, which can be encoded in, say, $cn/4$ bits assuming c is chosen properly, therefore the query string is in K.

In [Wat87], Watanabe used time-space bounded Kolmogorov complexity to construct a more sophisticated set D which is polynomial Turing complete for E but not complete for E under polynomial truth-table reduction. Allender and Watanabe [AW88] used Kolmogorov complexity to characterized the class of sets which are polynomial many-one equivalent to tally sets, in order to study the question of whether $E_m^P(Tally) = E_{btt}^P(Tally)$ is true, where $E_-^P(Tally) = \{L : \text{for some tally set } T, L =_-^P T\}$. In [Huy85,Huy86], Huynh started a series of studies on the concept of resource-bounded Kolmogorov complexity of languages. He defined the (time-/space-bounded) Kolmogorov complexity of a language to be the (time-/space-bounded) Kolmogorov complexity of $Seq(L^{<n}) = C_L(w_1)C_L(w_2)\cdots C_L(w_{2^n-1})$, where w_i is the lexicographically the ith word and $C_L(w_i) = 1$ if and only if $w_i \in L$. In particular, he shows that there is a language $L \in DTIME(2^{2^{O(n)}})$ (any hard set for this class) such that the 2^{poly}-time-bounded Kolmogorov complexity of L is exponential almost everywhere. That is, the sequence $Seq(L^{<n})$ cannot be compressed to a subexponentially short string within 2^{poly} time for all but finitely many n's. Similar results were also obtained for space-bounded classes. He used these results to classify exponential-size circuits.

Allender and Rubinstein studied the relation between small resource bounded Kolmogorov complexity and P-printability. Sets like $K[k \log n, n^k, \infty]$ for some constant k are said to have small time bounded Kolmogorov complexity. A set S is said to be polynomial-time printable (P-printable) if there exists a k such that all the elements of S up to size n can be printed by a deterministic machine in time $n^k + k$. Clearly every P-printable set is a sparse set in P. Define the *ranking function* for a language L, $r_L : \Sigma^* \to N$, is given by $r_L(x) = d(\{w \in L | w < x\})$ [GS85]. Allender and Rubinstein [AR88,BB86,HH86] proved

Theorem 7.4.17 *The following are equivalent:*

(1) S is P-printable.

(2) S is sparse and has a ranking function computable in polynomial time.

(3) S is P-isomorphic to some tally set in P.

(4) $S \subseteq K[k \log n, n^k]$ for some constant k and $S \in P$.

Note: Balcazar and Book, and Hartmanis and Hemachandra proved the equivalence of (1) and (4). Allender and Rubinstein proved all equivalences.
    Proof.

(1) $\to$ (2): immediate.

(2) $\to$ (3): Let S have a polynomial time computable ranking function r_1 and $d(S^{\leq n}) < p(n)$ for some polynomial $p(n)$. Then $r_2(w) = 1w - r_1(w)$ is a ranking function for the complement of S, where $1w$ is treated as a binary number. Also the set $T = \{0^{np(n)+i} | r_1(1^{n-1}) < i \leq r_1(1^n)\}$ is a tally set in P. Let r_3 be a ranking function for the complement of T. As was noted in [GS85], all these ranking functions have inverses which are computable in time polynomial in the length of their output. It is now easy to see that the function that maps x

of length n to $0^{np(n)+r_1(x)}$ if $x \in S$ and to $r_3^{-1}(r_2(x))$ if $x \notin S$ is a P-isomorphism which maps S onto T.

(3) $\to$ (4): By assumption, there is a P-isomorphism f, both f and f^{-1} computable in n^c for some c, which maps S onto a tally set, $T \subseteq \{0\}^*$, in P. Trivially, $S \in$ P. Since f is computable in time n^c, $|f(x)| \leq n^c$ for $l(x) = n$. Hence the binary representation of $f(x)$ has length $\leq c \log n$. Hence x can be represented by $f(x)$ in binary using only $c \log n$ bits. To compute x from $f(x)$, we simply compute $f^{-1}(0^{f(x)})$ which takes polynomial time by assumption. Thus $S \subseteq K[k \log n, n^k]$ for $k > c^2$.

(4) $\to$ (1): Assume that $S \in$ P and that for some k, $S \subseteq K_U[k \log n, n^k]$. On input n, for each of the $n^{k+1} - 1$ strings of length $\leq k \log n$, run the universal machine U for at most n^k steps and, if the computation has completed and the output of U is in S, print it. This process can clearly be done in time polynomial in n.

$\square$

Corollary 7.4.18 *There is a k such that $A \subseteq K[k \log n, n^k]$ if and only if A is P-isomorphic to a tally set.*

7.4.5   A KOLMOGOROV RANDOM REDUCTION

The original ideas of this Section belong to U. Vazirani and V. Vazirani [VV82]. We re-formulate their results in terms of Kolmogorov complexity.

In 1979 Adleman and Manders defined a probabilistic reduction, called UR-reduction, and showed several number-theoretic problems to be hard for NP under UR-reductions but not known to be NP-hard. In [VV82] the notion is refined as follows:

> A is PR-reducible to B, denoted as $A \leq_{\text{PR}} B$, if and only if there is a probabilistic polynomial time TM T and $\delta > 0$ such that (1) $x \in A$ implies $T(x) \in B$, and (2) x not in A implies $\text{Prob}(T(x) \text{ not in } B) \geq \delta$. A problem is PR-*complete* if every NP problem can be PR-reduced to it.

Vazirani and Vazirani obtained the first non number-theoretic PR-*complete* problem, which is still not known to be NP-complete, ENCODING BY TM:

INSTANCE: Two strings $x, y \in \{0, 1, 2, \alpha, \beta\}^*$, integer k.

QUESTION: Is there a TM M with k or fewer states that on input x generates y in $|y|$ steps. (M has one read-write tape initially containing x and a write-only tape to write y. M must write one symbol of y each step; i.e., real-time.)

PR-completeness Proof. We reduce ENCODING BY FST to our problem, where the former is NP-complete and is defined as:

INSTANCE: Two strings $x, y \in \{0, 1, 2\}^*$, $|x| = |y|$, and integer k.

QUESTION: Is there a finite state transducer M with k or less states that outputs y on input x. (Each step, M must read a symbol and output a symbol.)

Reduction: any instance (x, y, k) of ENCODING BY FST is transformed to (xr, yr, k) for ENCODING BY TM, where $K(r|x, y) \geq |r| - c_\delta$, and $r \in \{\alpha, \beta\}^*$. For a given δ, Prob(generate such an r) $\geq \delta$. Clearly if there is a FST F of at most k state that outputs y on input x, then we can construct a TM with outputs yr on input xr by simply adding two new transitions from each state back to itself on α, β and output what it reads. If there is no such FST, then the k state TM must reverse or stop its read head on prefix x when producing y. Hence it produces r without seeing r (notice the real-time requirement). Hence $K(r|x, y) = O(1)$, a contradiction.   □

7.5   Time Bounded Kolmogorov Complexity and Language Compression

If A is a recursive set and x is lexicographically the ith element in A, then we know $K(x) \leq \log i + c_A$ for some constant c_A not depending on x. In this section let us write $K^t(s)$ to denote the t-time bounded Kolmogorov complexity of s, and A^n to denote the set of elements in A of length n. Further let $A \in$ P; i.e., it can be decided in $p(n)$ time whether $x \in A$, where $n = |x|$ and p a polynomial. It is seductive to think that:

Conjecture 7.5.1

$$\exists c \; \forall s \in A^n \; [K^p(s) \leq \log d(A^n) + c_A],$$

where p is some other polynomial.

However in polynomial time a Turing machine cannot search through 2^n strings as was assumed in the recursive case. Whether or not the above conjecture is true is still an important open problem in time-bounded Kolmogorov complexity which we deal with in this section. It also has important consequences in language compression.

Definition 7.5.2 (Goldberg and Sipser) *[GS85]*

(1) *A function $f : \Sigma^* \to \Sigma^*$ is a compression of language L if f is one-to-one on L and for all except finitely many $x \in L$, $|f(x)| < |x|$.*

(2) *A language L is* compressible in time T *if there is a compression function f for L which can be computed in time T, and the inverse f^{-1} of f with domain $f(L)$, such that for any $x \in L$, $f^{-1}(f(x)) = x$, can also be computed in time T.*

(3) *Compression function f* optimally compresses *a language L if for any $x \in L$ of length n, $|f(x)| \leq \lceil \log(\sum_{i=0}^{n} d(L^i)) \rceil$.*

(4) *One natural and optimal compression is* ranking. *The ranking function $r_L : L \to N$ maps $x \in L$ to its index in a lexicographical ordering of L.*

Obviously, language compression is closely related to the Kolmogorov complexity of the elements in the language. *Efficient* language compression is closely related to the *time-bounded* Kolmogorov complexity of the elements of the language. By using a ranking function, we can obtain the optimal Kolmogorov complexity of any element in a recursive set, and hence, optimally compress the recursive set. That was trivial. Our purpose is to study the polynomial time setting of the problem. This is far from trivial.

7.5.1   WITH THE HELP OF AN ORACLE

Let $\psi_1, \psi_2, \ldots$ be an effective enumeration of partial recursive predicates. Let T_ψ be the first multitape Turing machine which computes ψ. $T_\psi(x)$ outputs 0 or 1. If T_ψ accepts x in t steps (time), then we also write $\psi^t(x) = 1$.

Definition 7.5.3 (Sipser) *Let x, y, p be strings in $\{0,1\}^*$. Fixing ψ, we can define KD_ψ^t of x, conditional to ψ and y by*

$$KD_\psi^t(x|y) = min\{|p| : \forall v, \psi^t(v, p, y) = 1 \text{ iff } v = x\},$$

and $KD^{t\psi}(x|y) = \infty$ if there are no such p. Define the unconditional resource bounded Kolmogorov complexity, of x as $K^t(x) = K^t(x|\epsilon)$.

Remark. One can prove an invariance theorem similar to that of K^t version, we hence can drop the index ψ in KD_ψ^t.

The intuition of the above definition is that while $K^t(x)$ is the length of the shortest program *generating* x in $t(|x|)$ time, $KD^t(x)$ is the length of the shortest program *accepting* only x in $t(|x|)$ time. In pure Kolmogorov complexity, these two measures differ by only an additive constant. In the resource bounded Kolmogorov complexity, they appear to be quite different. The KD version appears to be the simpler one, viz., Sipser proved [Sip83]: Let p, q be polynomials, c be a constant, and NP be an NP-complete oracle,

(1) $\forall p \exists q [KD^q(s) \leq K^p(s) + O(1)];$

(2) $\forall p \exists q [K^q(s|\text{NP}) \leq KD^p(s) + O(1)]$.

(3) $\forall c \exists d$, if $A \subseteq \sum^n$ and A is accepted by a circuit of size n^c, then $\forall s \in A$:

$$KD^d(s|A, i_A) \leq \log d(A) + \log \log d(A) + O(1).$$

where i_A depends on A and has length about $n \cdot \log d(A)$.

(4) $\forall c \exists d$, if $A \subseteq \sum^n$ is accepted by a circuit of size n^c and there is a string i_A such that for each $s \in A$,

$$K^d(s|A, i_A, \text{NP}) \leq \log d(A) + \log \log d(A) + O(1)$$

then

$$K^d(s|A, \Sigma_2^{\text{P}}) \leq \log d(A) + \log \log d(A) + O(1)$$

In order to prove the above results, Sipser needed an important coding lemma for which we present a new proof using Kolmogorov complexity. Let $A \subseteq \Sigma^n$, $k = d(A)$ and $m = 1 + \lceil \log k \rceil$. Let $h : \Sigma^n \to \Sigma^m$ be a linear transformation given by a randomly chosen $m \times n$ binary matrix $R = \{r_{ij}\}$; i.e., for $x \in \Sigma^n$, xR is a string $y \in \Sigma^m$ where $y_i = (\sum_j r_{ij} \times x_j) \bmod 2$. Let H be a collection of such functions. let $A, B \subseteq \Sigma^n$ and $x \in \Sigma^n$. We say h *separates* x *within* A if for every $y \in A$, different from x, $h(y) \neq h(x)$. h *separates* B *within* A if it separates each $x \in B$ within A. H *separates* B *within* A if for each $x \in B$ some $h \in H$ separates x within A. In order to give each element in A a (logarithmic) short code, we randomly hash elements of A into short codes. If collision can be avoided, then elements of A can be described by short programs.

Lemma 7.5.4 (Coding Lemma, Sipser) *Let $A \subseteq \Sigma^n$, where $d(A) = k$. Let $m = 2 + \lceil \log k \rceil$. There is a collection H of m random linear transformations $\Sigma^n \to \Sigma^m$, H separates A within A.*

Proof. Fix a random string s of length nm^2 such that $K(s|A) \geq |s|$. Cut x into m equal pieces. Use the nm bits from each piece to form a $n \times m$ binary matrices in the obvious way. Thus we have constructed a set H of m random matrices. We claim that H separates A within A.

Assume this is not true. That is, for some $x \in A$, no $h \in H$ separates x within A. Hence there exist $y_1, \ldots, y_m \in A$ such that $h_i(x) = h_i(y_i)$. Hence $h_i(x - y_i) = 0$. Since $x - y_i \neq 0$, the first column of h_i corresponding to a 1 in $x - y_i$ can be expressed by the rest of the columns using $x - y_i$. Now we can describe s using the following

- this discussion;

- index of x in A, using $\lceil \log k \rceil$ bits;

- indices of $y_1, \ldots, y_m$, in at most $m\lceil \log k \rceil$ bits,

- matrices $h_1, \ldots, h_m$ each minus the redundant column, in $m^2n - m^2$ bits.

From the above information, a short program, given A, will reconstruct h_i by using the rest of the columns of h_i and x, y_i. The total length is only

$$m^2n - m(\log k + 1) + \log k + m(\log k) + O(1) \leq nm^2 - \Omega(m).$$

Hence, $K(s|A) < |s|$, a contradiction. $\qquad\qquad\qquad\qquad\qquad\square$

From this lemma, Sipser also proved BPP $\subseteq \Sigma_4^P \cap \Pi_4^P$. Gács improved this to

Theorem 7.5.5 BPP $\subseteq \Sigma_2^P \cap \Pi_2^P$.

Proof (Gács). Let $B \in$ BPP be accepted by a probabilistic algorithm with error probability at most 2^{-n} on inputs of length n, which uses $m = n^k$ random bits. Let $E_x \subset \Sigma^m$ be the collection of random inputs on which M rejects x. For $x \in B$, $|E_x| \leq 2^{m-n}$. Letting $l = 1 + m - n$, the Sipser's Coding Lemma states that there is a collection H of l linear transformations from Σ^m to Σ^l separating E_x within E_x. If x is not in B, $|E_x| > 2^{m-1}$ and by the pigeon hole principle, no such collection exists. Hence $x \in B$ if and only if such an H exists. The latter can be expressed as

$$\exists H \; \forall e \in E_x \; \exists h \in H \; \forall e' \in E_x \; [e \neq e' \Rightarrow h(e) \neq h(e')]$$

The second existential quantifier has polynomial range hence can be eliminated. Hence BPP $\in \Sigma_2^P$. Since BPP is closed under complement, BPP $\in \Pi_2^P$. Hence BPP $\in \Sigma_2^P \cap \Pi_2^P$. $\qquad\qquad\qquad\square$

7.5.2 LANGUAGE COMPRESSION WITHOUT ORACLE

Without the help of oracles, Sipser and Goldberg [GS85] obtained much weaker and more difficult results. For a given language L, define the density of L to be $\mu_L = max\{\mu_L(n)\}$, where $\mu_L(n) = d(L^n)/2^n$. Goldberg and Sipser proved: If $L \in$ P, $k > 3$, and $\mu_L \leq n^{-k}$, then L can be compressed in probabilistic polynomial time; the compression function f maps strings of length n to strings of length $n - (k - 3)\log n + c$.

The above result is weak in two senses. First, if a language L is very sparse, say $\mu_L \leq 2^{-n/2}$, then one expects to compress $n/2$ bits instead of only $O(\log n)$ bits given by the theorem. Can this be improved? Second, the current compression algorithm is probabilistic. Can this be made deterministic? In computational complexity, oracles sometimes help us to understand the possibility of proving a new theorem. Goldberg and Sipser show that when S, the language to be compressed, does not have to be in P and the membership query of S is given by an oracle, then the above result is optimal. Specifically:

(1) There is a sparse language S which cannot be compressed by more than $O(\log n)$ bits by a probabilistic polynomial time machine with an oracle for S.

(2) There is a language S, of density $\mu_S < 2^{-n/2}$, which cannot be compressed by any deterministic polynomial time machine that uses the oracle for S.

See [Sto88] for practical data compression techniques.

7.5.3   RANKING: OPTIMALLY COMPRESSIBLE LANGUAGES

Ranking is a special and optimal case of compression. The ranking function r_L maps the strings in L to their indices in the lexicographical ordering of L. If $r_L : L \to N$ is polynomial time computable, then so is $r_L^{-1} : N \to L$. We are only interested in polynomial time computable ranking functions. In fact there are natural language classes that are easy to compress. Goldberg and Sipser [GS85] and Allender [All85] show: If a language L is accepted by a one-way log space Turing machine, then r_L can be computed in polynomial time. Goldberg and Sipser also prove by diagonalization: that (a) there is an exponential time language that cannot be compressed in deterministic polynomial time; and (b) there is a double exponential time language that cannot be compressed in probabilistic polynomial time. Call **C** P-rankable if for all $L \in$ **C**, r_L is polynomial time computable. Hemachandra in [Hem87] proved that P is P-rankable if and only if NP is P-rankable, and P is P-rankable if and only if $P = P^{\#P}$, and PSPACE is P-rankable if and only if $P = PSPACE$. Say a set A is k-enumeratively-rankable if there is a polynomial time computable function f so that for every x, $f(x)$ prints a set of k numbers, one of which is the rank of x with respect to A. Cai and Hemachandra [CH86] proved $P = P^{\#P}$ if and only if each set $A \in P$ for some k_A has a k_A-enumerative-ranker.

7.6   Conclusion

The opinion has sometimes been voiced that Kolmogorov complexity has only very abstract use. We are convinced that Kolmogorov complexity is immensely useful in a plethora of applications in the study of computational complexity. We believe that we have given conclusive evidence for that conviction by this collection of applications.

In our view the covered material represents only the onset of a potentially enormous number of applications of Kolmogorov complexity in computer sciences. If, by the examples we have discussed, readers get the feel of how to use this general purpose tool in their own applications, this exposition would have served its purpose.

Acknowledgements: We are deeply in debt to Juris Hartmanis and Joel Seiferas who have introduced us in various ways to Kolmogorov complexity. We are grateful to Greg Chaitin, Peter Gács, Leonid Levin and Ray Solomonoff for taking lots of time to tell us about the early history of our subject, and introducing us to many exciting applications. Additional comments were provided by Donald Loveland and Albert Meyer. R. Paturi, J. Seiferas and Y.Yesha, kindly supplied to us (and gave us permission to use) their unpublished material about lower bounds for probabilistic machines and Boolean matrix rank, respectively. L. Longpré, B.K. Natarajan, and O. Watanabe supplied to us interesting material in connection with the structure of complexity classes. Comments of Charles Bennett, Peter van Emde Boas, Jan Heering, Evangelos Kranakis, Ker-I Ko, Danny Krizanc, Michiel van Lambalgen, Lambert Meertens, A. Kh. Shen', Umesh Vazirani, and A. Verashagin are gratefully acknowledged. John Trump read the manuscript carefully and discovered several errors. He is not responsible for the remaining ones. We also thank Sarah Mocas for her help. Th. Tsantilas and J. Rothstein supplied many useful references. The Chairman of Computer Science Department at York University, Eshrat Arjomandi, provided much support during this work.

7.7 REFERENCES

[Aan74]   S.O. Aanderaa. On k-tape versus (k-1)-tape real-time computation. In R.M. Karp, editor, *Complexity of Computation*, pages 75–96, American Math. Society, Providence, R.I., 1974.

[Adl79]   L. Adleman. *Time, space, and randomness*. Technical Report MIT/LCS/79/TM-131, Massachusetts Institute of Technology, Laboratory for Computer Science, March 1979.

[All85]   E. Allender. *Invertible Functions*. PhD thesis, Georgia Institute of Technology, 1985.

[All87]   E. Allender. Some consequences of the existence of pseudorandom generators. In *Proc. 19th ACM Symposium on Theory of Computation*, pages 151–159, 1987.

[AR88]   E.A. Allender and R.S. Rubinstein. P-printable sets. *SIAM J. on Computing*, 17, 1988.

[AW88]   E. Allender and O. Watanabe. Kolmogorov complexity and de-
          grees of tally sets. In *Proceedings 3rd Conference on Structure
          in Complexity Theory*, pages 102–111, 1988.

[Bar68]   Y.M. Barzdin'. Complexity of programs to determine whether
          natural numbers not greater than n belong to a recursively
          enumerable set. *Soviet Math. Dokl.*, 9:1251–1254, 1968.

[BB86]   J. Balcazar and R. Book. On generalized Kolmogorov complex-
          ity. In *Proc. 1st Structure in Complexity Theory Conference,
          Lecture Notes in Computer Science, vol. 223*, pages 334–340,
          Springer Verlag, Berlin, 1986.

[BC80]   A. Borodin and S. Cook. A time-space tradeoff for sorting on a
          general sequential model of computation. In *12th ACM Symp.
          on Theory of Computing*, 1980.

[Ben87]   C.H. Bennett. Dissipation, information, computational com-
          plexity and the definition of organization. In D. Pines, editor,
          Emerging Syntheses in Science, Addison-Wesley, Reading, MA,
          1987. (Proceedings of the Founding Workshops of the Santa Fe
          Institute, 1985, pages 297-313.).

[Ben88]   C.H. Bennett. On the logical "depth" of sequences and their
          reducibilities to random sequences. In R. Herken, editor, *The
          Universal Turing Machine - A Half-Century Survey*, pages 227–
          258, Oxford University Press, 1988.

[BFK*79] A. Borodin, M.J. Fischer, D.G. Kirkpatrick, N.A. Lynch, and
          M. Tompa. A time-space tradeoff for sorting and related non-
          oblivious computations. In *20th IEEE Symposium on Founda-
          tions of Computer Science*, pages 319–328, 1979.

[BH77]   L. Berman and J. Hartmanis. On isomorphisms and density of
          NP and other complete sets. *SIAM J. on Computing*, 6:305–
          327, 1977.

[BORW88] R. Book, P. Orponen, D. Russo, and O. Watanabe. Lowness
          properties of sets in the exponential-time hierarchy. *SIAM J.
          Computing*, 17:504–516, 1988.

[CH86]   J. Cai and L. Hemachandra. *Exact counting is as easy as ap-
          proximate counting*. Technical Report 86-761, Computer Sci-
          ence Department, Cornell University, Ithaca, N.Y., 1986.

[Cha69]   G.J. Chaitin. On the length of programs for computing finite
          binary sequences: statistical considerations. *J. Assoc. Comp.
          Mach.*, 16:145–159, 1969.

[Cha77]   G.J. Chaitin. Algorithmic Information Theory. *IBM J. Res. Dev.*, 21:350–359, 1977.

[Chr86]   M. Chrobak. Hierarchies of one-way multihead automata languages. *Theoretical Computer Science*, 48:153–181, 1986. (also in 12th ICALP, Springer LNCS 194:101-110, 1985).

[CL86]   M. Chrobak and M. Li. k+1 heads are better than k for PDAs. In *Proc. 27th IEEE Symposium on Foundations of Computer Science*, pages 361–367, 1986.

[Cuy84]   R.R. Cuykendall. *Kolmogorov information and VLSI lower bounds*. PhD thesis, University of California, Los Angeles, Dec. 1984.

[CW79]   J. Carter and M. Wegman. Universal classes of hashing functions. *J. Comp. Syst. Sciences*, 18:143–154, 1979.

[Dal77]   R.P. Daley. On the inference of optimal descriptions. *Theoretical Computer Science*, 4:301–309, 1977.

[DGPR84]   P. Duris, Z. Galil, W. Paul, and R. Reischuk. Two nonlinear lower bounds for on-line computations. *Information and Control*, 60:1–11, 1984.

[Die87]   M. Dietzfelbinger. *Lower bounds on computation time for various models in computational complexity theory*. PhD thesis, University of Illinois at Chicago, 1987.

[ES74]   P. Erdös and J. Spencer. *Probabilistic methods in combinatorics*. Academic Press, New York, 1974.

[Flo68]   R. Floyd. Review 14. *Comput. Rev.*, 9:280, 1968.

[Gal69]   H. Gallaire. Recognition time of context-free languages by online Turing machines. *Information and Control*, 15:288–295, 1969.

[GKS86]   Z. Galil, R. Kannan, and E. Szemeredi. On nontrivial separators for k-page graphs and simulations by nondeterministic one-tape Turing machines. In *Proc. 18th ACM Symp. on Theory of Computing*, pages 39–49, 1986.

[GL]   M. Gereb and M. Li. Lower bounds in string matching. In preparation.

[GS81]   Z. Galil and J. Seiferas. Time-space optimal matching. In *Proc. 13th ACM Symposium on Theory of Computing*, 1981.

198   Ming Li, Paul M.B. Vitányi

[GS85]   Y. Goldberg and M. Sipser. Compression and Ranking. In
            *Proc. 17th Assoc. Comp. Mach. Symposium on Theory of Com-
            puting*, pages 440–448, 1985.

[Har83]   J. Hartmanis. Generalized Kolmogorov complexity and the
            structure of feasible computations. In *Proc. 24th IEEE Sym-
            posium on Foundations of Computer Science*, pages 439–445,
            1983.

[Hem86]   L. Hemachandra. *Can P and NP manufacture randomness?*
            Technical Report 86-795, Computer Science Department, Cor-
            nell University, Ithaca, N.Y., December 1986.

[Hem87]   L. Hemachandra. On ranking. In *Proc. 2nd Ann. IEEE Con-
            ference on Structure in Complexity Theory*, pages 103–117,
            1987.

[HH86]   J. Hartmanis and L. Hemachandra. On sparse oracles sep-
            arating feasible complexity classes. In *Proc. 3rd Symposium
            on Theoretical Aspects of Computer Science, Lecture Notes in
            Computer Science, vol. 210*, pages 321–333, Springer Verlag,
            Berlin, 1986.

[HI68]   M.A. Harrison and O.H. Ibarra. Multi-head and multi-tape
            pushdown automata. *Information and Control*, 13:433–470,
            1968.

[HU79]   J.E. Hopcroft and J.D. Ullman. *Introduction to Automata The-
            ory, Languages, and Computation.* Addison-Wesley, 1979.

[Huy85]   D.T. Huynh. *Non-uniform complexity and the randomness of
            certian complete languages.* Technical Report 85-34, Iowa State
            University, December 1985.

[Huy86]   D.T. Huynh. Resource-bounded Kolmogorov complexity of
            hard languages. In *Proc. 1st Structure in Complexity The-
            ory Conference, Lecture Notes in Computer Science, vol. 223*,
            pages 184–195, Springer Verlag, Berlin, 1986.

[HW85]   F. Meyer auf der Heide and A. Wigderson. The complexity
            of parallel sorting. In *Proc. 17th ACM Symp. on Theory of
            computing*, pages 532–540, 1985.

[IK75]   O.H. Ibarra and C.E. Kim. On 3-head versus 2 head finite
            automata. *Acta Infomatica*, 4:193–200, 1975.

[IL89]   Y. Itkis and L.A. Levin. Power of fast VLSI models is insensitive to wires' thinnes. In *Proc. 30th IEEE Symposium on Foundations of Computing*, 1989.

[IM]   A. Israeli and S. Moran. Private communication.

[Ko83]   Ker-I Ko. *Resource-bounded program-size complexity and pseudorandom sequences*. Technical Report, Department of computer science, University of Houston, 1983.

[Kol65]   A.N. Kolmogorov. Three approaches to the quantitative definition of information. *Problems in Information Transmission*, 1:1–7, 1965.

[Lev73]   L.A. Levin. Universal search problems. *Problems in Information Transmission*, 9:265–266, 1973.

[Lev74]   L.A. Levin. Laws of information conservation (non-growth) and aspects of the foundation of probability theory. *Problems in Information Transmission*, 10:206–210, 1974.

[Lev83]   L.A. Levin. Do chips need wires? Manuscript/NSF proposal MCS-8304498, 1983. Computer Science Department, Boston University.

[Lev84]   L.A. Levin. Randomness conservation inequalities; information and independence in mathematical theories. *Information and Control*, 61:15–37, 1984.

[Li85a]   M. Li. *Lower Bounds in Computational Complexity, Report TR-85-663*. PhD thesis, Computer Science Department, Cornell University, March 1985.

[Li85b]   M. Li. Simulating two pushdowns by one tape in $O(n^{**}1.5$ $(\log n)^{**}0.5)$ time. In *Proc. 26th IEEE Symposium on the Foundations of Computer Science*, 1985.

[LLV86]   M. Li, L. Longpré, and P.M.B. Vitányi. On the power of the queue. In *Structure in Complexity Theory, Lecture Notes in Computer Science, volume 223*, pages 219–233, Springer Verlag, Berlin, 1986.

[Lon86]   L. Longpré. *Resource bounded Kolmogorov complexity, a link between computational complexity and information theory*, Techn. Rept. TR-86-776. PhD thesis, Computer Science Department, Cornell University, 1986.

[Lou83]   M. Loui. Optimal dynamic embedding of trees into arrays. *SIAM J. Computing*, 12:463–472, 1983.

[LS81]   R. Lipton and R. Sedgewick. Lower bounds for VLSI. In *13th ACM Symp. on Theory of computing*, pages 300–307, 1981.

[LVa]   M. Li and P.M.B. Vitányi. An Introduction to Kolmogorov Complexity and Its Applications. (To appear, Addison-Wesley).

[LVb]   M. Li and P.M.B. Vitányi. Kolmogorov Complexity and its Applications. J. van Leeuwen, editor, Handbook of Theoretical Computer Science, North-Holland. To appear.

[LV88a]   M. Li and P.M.B. Vitányi. Kolmogorovskaya slozhnost' dvadsat' let spustia. *Uspekhi Mat. Nauk*, 43:6:129–166, 1988. (In Russian; = *Russian Mathematical Surveys*).

[LV88b]   M. Li and P.M.B. Vitányi. Tape versus queue and stacks: the lower bounds. *Information and Computation*, 78:56–85, 1988.

[LV89a]   M. Li and P.M.B. Vitányi. Inductive reasoning and Kolmogorov complexity. In *Proc. 4th IEEE Conference on Structure in Complexity Theory*, pages 165–185, 1989.

[LV89b]   M. Li and P.M.B. Vitányi. A new approach to formal language theory by Kolmogorov complexity. In *16th International Conference on Automata, Languages and Programming, Lecture Notes in Computer Science, vol. 372*, pages 488–505, Springer Verlag, Berlin, 1989.

[LY86a]   M. Li and Y. Yesha. New lower bounds for parallel computation. In *Proc. 18th Assoc. Comp. Mach. Symposium on Theory of Computing*, pages 177–187, 1986.

[LY86b]   M. Li and Y. Yesha. String-matching cannot be done by 2-head 1-way deterministic finite automata. *Information Processing Letters*, 22:231–235, 1986.

[Maa85]   W. Maass. Combinatorial lower bound arguments for deterministic and nondeterministic Turing machines. *Trans. Amer. Math. Soc.*, 292:675–693, 1985.

[Mai83]   H.G. Mairson. The program complexity of searching a table. In *Proceedings 24th IEEE Symposium on Fundations of Computer Science*, pages 40–47, 1983.

[Meh82]   K. Mehlhorn. On the program-size of perfect and universal hash functions. In *Proc. 23rd Ann. IEEE Symposium on Foundations of Computer Science*, pages 170–175, 1982.

[Miy82]   S. Miyano. A hierarchy theorem for multihead stack-counter automata. *Acta Informatica*, 17:63–67, 1982.

[Miy83]   S. Miyano. Remarks on multihead pushdown automata and multihead stack automata. *Journal of Computer and System Sciences*, 27:116–124, 1983.

[MS86]   W. Maass and G. Schnitger. An optimal lower bound for Turing machines with one work tape and a two-way input tape. In *Structure in Complexity Theory, Lecture Notes in Computer Science, volume 223*, pages 249–264, Springer Verlag, Berlin, 1986.

[MSS87]   W. Maass, G. Schnitger, and E. Szemeredi. Two tapes are better than one for off-line Turing machines. In *Proc. 19th ACM Symposium on Theory of Computing*, pages 94–100, 1987.

[Nat88]   B. K. Natarajan. 1988. Personal communication.

[Nel76]   C.G. Nelson. *One-way automata on bounded languages*. Technical Report TR14-76, Harvard University, July 1976.

[Par84]   I. Parberry. *A complexity theory of parallel computation*. PhD thesis, Warwick University, 1984.

[Pau79]   W. Paul. Kolmogorov's complexity and lower bounds. In *Proc. 2nd International Conference on Fundamentals of Computation Theory*, September 1979.

[Pau82]   W.J. Paul. On-line simulation of k+1 tapes by k tapes requires nonlinear time. *Information and Control*, 1–8, 1982.

[Pau84]   W. Paul. On heads versus tapes. *Theoretical Computer Science*, 28:1–12, 1984.

[Pet80]   G. Peterson. Succinct representations, random strings and complexity classes. In *Proc. 21st IEEE Symposium on Foundations of Computer Science*, pages 86–95, 1980.

[PS83]   R. Paturi and J. Simon. Lower bounds on the time of probabilistic on-line simulations. In *Proc. 24th Ann. IEEE Symposium on Foundations of Computer Science*, page 343, 1983.

202   Ming Li, Paul M.B. Vitányi

[PSNS88]   R. Paturi, J. Simon, R.E. Newman-Wolfe, and J. Seiferas. Milking the Aanderaa argument. April 1988. University of Rochester, Comp. Sci. Dept., Rochester, N.Y.

[PSS81]   W.J. Paul, J.I. Seiferas, and J. Simon. An information theoretic approach to time bounds for on-line computation. *J. Computer and System Sciences*, 23:108–126, 1981.

[Rog67]   H. Rogers, Jr. *Theory of Recursive Functions and effective computability.* McGraw-Hill, New York, 1967.

[Ros65]   A. Rosenberg. *Nonwriting extensions of finite automata.* PhD thesis, Harvard University, 1965.

[Ros66]   A. Rosenberg. On multihead finite automata. *IBM J. Res. Develop.*, 10:388–394, 1966.

[RS82]   S. Reisch and G. Schnitger. Three applications of Kolmogorov-complexity. In *Proc. 23rd Ann. IEEE Symposium on Foundations of Computer Science*, pages 45–52, 1982.

[Sei85]   J. Seiferas. The symmetry of information, and An application of the symmetry of information. Notes, August 1985. Computer Science Dept, University of Rochester.

[Sei86]   J. Seiferas. A simplified lower bound for context-free-language recognition. *Information and Control*, 69:255–260, 1986.

[Sip83]   M. Sipser. A complexity theoretic approach to randomness. In *Proceedings 15th Assoc. Comp. Mach. Symposium on Theory of Computing*, pages 330–335, 1983.

[Sol64]   R. J. Solomonoff. A formal theory of inductive inference, Part 1 and Part 2. *Information and Control*, 7:1–22, 224–254, 1964.

[Sto88]   J. Storer. *Data Compression: Method and Theory*, chapter 6. Computer Science Press, Rockville, MD, 1988.

[Sud74]   I.H. Sudborough. *Computation by multi-head writing finite automata.* PhD thesis, Pennsylvania State University, University Park, 1974.

[Sud76]   I.H. Sudborough. One-way multihead writing finite automata. *Information and Control*, 30:1–20, 1976. (Also FOCS 1971).

[Tho79]   C.D. Thompson. Area-Time complexity for VLSI. In *Proc. 11th ACM Symp. on Theory of Computing*, pages 81–88, 1979.

[VB81]   L. Valiant and G. Brebner. Universal Schemes for parallel com-
            munication. In *Proc. 13th ACM Symp. on Theory of Comput-
            ing*, pages 263–277, 1981.

[Vit84a]   P.M.B. Vitányi. On the simulation of many storage heads by
            one. *Theoretical Computer Science*, 34:157–168, 1984. (Also,
            ICALP '83.).

[Vit84b]   P.M.B. Vitányi. On two-tape real-time computation and
            queues. *Journal of Computer and System Sciences*, 29:303–
            311, 1984.

[Vit85a]   P.M.B. Vitányi. An N**1.618 lower bound on the time to sim-
            ulate one queue or two pushdown stores by one tape. *Informa-
            tion Processing Letters*, 21:147–152, 1985.

[Vit85b]   P.M.B. Vitányi. Square time is optimal for the simulation of a
            pushdown store by an oblivious one-head tape unit. *Informa-
            tion Processing Letters*, 21:87–91, 1985.

[VV82]   U. Vazirani and V. Vazirani. A natural encoding scheme proved
            probabilistic polynomial complete. In *Proc. 23rd IEEE Symp.
            on Foundations of Computer Science*, pages 40–44, 1982.

[Wat87]   O. Watanabe. Comparison of polynomial time completeness
            notions. *Theoretical Computer Science*, 53, 1987.

[Yao82]   A. Yao. Theory and application of trapdoor functions. In *Pro-
            ceedings 23rd IEEE Symposium on Foundations of Computer
            Science*, pages 80–91, 1982.

[Yes84]   Y. Yesha. Time-space tradeoffs for matrix multiplication and
            discrete Fourier transform on any general random access com-
            puter. *J. Comput. Syst. Sciences*, 29:183–197, 1984.

[YR78]   A.C.-C. Yao and R.L. Rivest. k+1 heads are better than k. *J.
            Assoc. Comput. Mach.*, 25:337–340, 1978. (also see Proc. 17th
            FOCS, pages 67–70, 1976).

[ZL70]   A.K. Zvonkin and L.A. Levin. The complexity of finite objects
            and the development of the concepts of information and ran-
            domness by means of the Theory of Algorithms. *Russ. Math.
            Surv.*, 25:83–124, 1970.

8

The Power of Counting

Uwe Schöning[1]

ABSTRACT In this overview, various applications and variations of counting in structural complexity theory are discussed. The ability of exact counting is shown to be closely related with the ability of nondeterministic complementation. Relations between counting classes and classes requiring unique or few accepting computations are revealed. Further, approximate counting and relativized results are discussed.

8.1 Counting And Complementation

The recent breakthrough in showing that all nondeterministic space complexity classes are closed under complementation [Im88,Sz87] drastically demonstrated that the ability of exact counting – in this case, the number of reachable configurations – makes it possible to complement nondeterministically.

We assume the reader is familiar with Immerman's argument: The first part of his proof (Lemma 1 in [Im88]) is that, given the input x and additionally (as "side information" or "advice"), the number k of configurations reachable from the start configuration on input x, it is possible to present a nondeterministic machine which accepts if and only if x is not in the language. The second part of the proof (Lemma 2 in [Im88]) shows how to calculate k nondeterministically by an inductive counting argument.

It is interesting to note that similar arguments to Lemma 1 based on the census of the language instead of the number of reachable configurations appeared in the literature already before in a variety of contexts. (Probably the first use of this technique can be found in [Ba68].)

For a language A define $cens_A : \mathbb{N} \to \mathbb{N}$ such that $cens_A(n)$ is the number of strings in A of size n. Furthermore, call a set A *sparse* if $cens_A(n) \leq p(n)$ for some polynomial p and all n.

Given a language A, consider the following nondeterministic algorithm.

input (x, k) ;

---

[1]Universität Ulm, Abt. Theoretische Informatik, Postfach 4066, D-7900 Ulm, F.R.G.

guess a sequence of k different strings $y_1, \ldots, y_k$;
verify that $y_1, \ldots, y_k \in A$;
verify that $y_1, \ldots, y_k \neq x$;
accept if the verifications succeed.

If $k = cens_A(|x|)$, then it is clear that the algorithm accepts (x, k) if and only if x is not in A. Therefore, we get the following observation.

Proposition 8.1.1 ([MH80]) *If A is a sparse set in* NP *with $cens_A$ computable in polynomial time, then $\overline{A} \in$ NP. Therefore, if such a set would be NP-complete, then* NP $=$ co–NP.

This observation was an intermediate step in Mahaney's ultimate proof [Ma82] that no sparse set can be NP-complete unless P $=$ NP. As a strengthening of the above, observe that it suffices that $cens_A$ is computable *nondeterministically* – like the number of reachable configurations in Immerman's proof.

The census information in the above algorithm comes as "side information" or "advice". It has to be calculated separately, or just assumed to be given "for free". Karp and Lipton [KL80] define a model to deal with such advice information. For a class of sets C and a class of functions $\mathcal{F}$ from $\mathbb{N}$ to Σ^*, define $C/\mathcal{F}$ as the class of languages A for which there is a $B \in C$ and a $f \in \mathcal{F}$ such that for all $n \in \mathbb{N}$:

$$A = \{x \mid (x, f(|x|)) \in B\}$$

Here, $(\cdot, \cdot)$ is some standard pairing function on Σ^*. Let log, lin, poly, exp denote special instances of such classes $\mathcal{F}$, where $f \in$ log (lin, poly, exp) if $|f(n)|$ can be bounded by a logarithmic (linear, polynomial, exponential) function.

It is clear that every set A (recursive or not) is in P/exp. Furthermore, the class P/poly can be shown to coincide exactly with the class of sets having "polynomial size circuits". Another characterization of this class is as the class of sets which can be recognized by deterministic, polynomial time oracle machines with some sparse oracle set. In symbols,

$$\text{P/poly} = \{A \mid A \in \text{P}^S \text{ for some sparse set } S\}.$$

The following is a selection of Karp and Lipton's main results.

- If every set in NP is in P/log, then P $=$ NP.

- If every set in NP is in P/poly, then the polynomial hierarchy PH collapses to Σ_2^P. (The polynomial hierarchy is defined in the next section.)

- If every set in E = DTIME($2^{O(n)}$) is in P/poly, then E = PH = Σ_2^P, and additionally, P $\neq$ NP.

Note that for every set A, $bin(cens_A) \in$ lin, and for every sparse set A, $bin(cens_A) \in$ log. Here, $bin(k)$ is the binary expansion of number k. Therefore, the following modification of Proposition 8.1.1 is straightforward.

Proposition 8.1.2   (a) For every sparse set A in NP, $\overline{A} \in$ NP/log.

(b) For every sparse set A in NE, $\overline{A} \in$ NE/log.

(c) For every set A in NE, $\overline{A} \in$ NE/lin.

Pushing these ideas a little further, Kadin [Ka87] proved the following result, hereby improving a theorem of Mahaney [Ma82].

Theorem 8.1.3 ([Ka87]) If $A \in$ NP^S for some sparse set $S \in$ NP, then $A \in$ P^{NP}[log]. (Here, the suffix [log] indicates a logarithmic upper bound for the number of queries of the oracle machine.)
Therefore, if co–NP $\subseteq$ NP$^{\{S \in NP \mid S \text{ is sparse}\}}$, then the polynomial hierarchy collapses to P^{NP}[log].

Proof: Let $A = L(M, S)$ for some nondeterministic, polynomial time oracle machine M and a sparse oracle set $S \in$ NP. Compute A by the following deterministic oracle algorithm.

>   input x ;
>   by binary search, find the maximum k such that $(x, k) \in B$;
>   accept if $(x, k) \in C$.

Here, the following two oracle sets are used:

$$B = \{(x, k) \mid \text{there are at least } k \text{ different strings of size } |x| \text{ in } S\}$$

and

$$C = \{(x, k) \mid M \text{ accepts } x \text{ if oracle queries for some string } w \text{ are handled as follows. Answer "yes" if } w \text{ is in } S. \text{ Answer "no" if there are } k \text{ different strings of size } |x| \text{ in } S, \text{ and all of them are different from } x\}.$$

The oracles B and C are in NP, therefore A is in P^{NP}. Further, by using binary search, only logarithmically many oracle queries are needed.   □

For a class of sets $\mathcal{C}$, denote by $\mathrm{P}_{tt}^{\mathcal{C}}$ the class of sets that can be reduced to a set in $\mathcal{C}$ by a polynomial-time truth-table reduction. (A truth-table reduction is the special case of an oracle Turing machine reduction where all oracle queries have to be asked in advance. The rest of the computation leading to acceptance or rejection can be thought of as evaluating a truth-table depending on the oracle answers. Such an oracle access mechanism is sometimes referred to as parallel or nonadaptive as opposed to sequential or adaptive in the general case.)

The following theorem follows easily by adapting the proof of the previous theorem.

Theorem 8.1.4
$$\mathrm{P}_{tt}^{\mathrm{NP}} = \mathrm{P}^{\mathrm{NP}}[\log]$$

Proof: The backward inclusion is easy to see: An oracle computation tree with at most logarithmically many queries on each computation path does not have more than polynomially many queries on the whole tree. Therefore, these polynomially many queries can be asked in advance, thus giving a truth table reduction to the same set.

Conversely, given a $p(n)$ time-bounded truth-table reduction M from A to a set $S \in \mathrm{NP}$, we can proceed as in the proof of Theorem 8.1.3, but the definition of the NP-oracle sets B and C has to be altered.

$B = \{(x, k) \mid$ there are at least k different strings of size $\leq p(|x|)$ in S which are queried by M on input $x\}$

and

$C = \{(x, k) \mid M$ accepts x if oracle queries for some string w are handled as follows. Answer "yes" if w is in S. Answer "no" if there are k different strings of size $\leq p(|x|)$ in S, and all of them are queried by M on x, and all of them are different from $x\}$.

□

As in Proposition 8.1.2, stepping from NP to NE, we can do without sparseness requirements, and immediately obtain

Theorem 8.1.5 ([SW88]) *If $A \in \mathrm{NEXP}^{\mathrm{NEXP}}\{\mathrm{poly}\}$, then $A \in \mathrm{P}^{\mathrm{NEXP}}$. (Here, the suffix $\{\mathrm{poly}\}$ indicates a polynomial upper bound for the size of oracle queries.)*

A consequence is the collapse of the *strong exponential hierarchy*, originally proved by Hemachandra [He87a,He87c] (with a much more complicated construction):

Corollary 8.1.6

$$\mathrm{NP}^{\mathrm{NP}^{\cdots^{\mathrm{NP}^{\mathrm{NEXP}}}}} = \mathrm{P}^{\mathrm{NEXP}}$$

Discussion. We have seen that the ability of exact counting is closely related with the ability of nondeterministic complementation. The clever idea in [Im88,St77] is *not* to count the number of strings of size n, but the number of configurations reachable from the start configurations. The advantage is that the underlying *structure* can be used: each configuration has a finite number of successor configurations which can be calculated easily. On the set of strings of a given length no such structure seems to be available. Other such possibilities of counting and their subtle differences are discussed in a later section.

We conjecture that additional structural assumptions, such as *self-reducibility*, might lead to improved versions of the above mentioned results, maybe even to closure under complementation of certain classes. The following open problems do not seem out of reach:

- Are the sparse or tally sets in NP already in co–NP?

- Equivalently, does the strong exponential hierarchy not only collapse to its Δ_2 level, but to its Σ_1 level? In other words, is it true that NEXP = co–NEXP?

- An easier question is whether all sparse sets in NEXP are already in co–NEXP.

- How can self-reducibility help to improve the above statements?

8.2   Counting Quantifiers And Hierarchies

Quantifiers which assert the existence of a unique object, or the existence of a specific number of objects, have been known in Logic for a long time. Wagner [Wa86a,Wa86b] introduced this concept in the polynomial-time bounded setting as follows. For a class $\mathcal{K}$ let $C\mathcal{K}$ be the class of sets A such that there is a set B in $\mathcal{K}$, a polynomial p and a polynomial-time computable function t (the *threshold function*) such that

$$A = \{x \mid \text{there exist at least } t(x)\text{-many strings } y$$
$$\text{of size at most } p(|x|) \text{ such that } (x,y) \in B\}$$

Similarly, define the classes $\exists\mathcal{K}$ with "there exists at least one y", and $\forall\mathcal{K}$ with "for all y". It is clear the the classes Σ_1^P, Π_1^P, Σ_2^P, Π_2^P,... of the *polynomial-time hierarchy* [St77,Wr77] are exactly the classes $\exists$P, $\forall$P, $\exists\forall$P, $\forall\exists$P,....

By setting the threshold in the definition of $C\mathcal{K}$ to 1 or to the total number of y's, the following Proposition is immediate.

Proposition 8.2.1 *For all classes* $\mathcal{K}$, $\exists\mathcal{K} \cup \forall\mathcal{K} \subseteq C\mathcal{K}$.

By introducing dummy computation paths, the threshold can be changed without changing the defined class $C\mathcal{K}$. Therefore, it can be seen that the class CP equals Gill's class PP [Gi77] (which has also been named mNP in [Si77]). The definition of PP requires that the threshold function is fixed to be half of the number of potential y's.

It is not clear whether two counting quantifiers can be merged into one. So there might exist an infinite hierarchy of classes, the *polynomial-time counting hierarchy*, CP, CCP, CCCP, ... inside PSPACE. (It is clear that all these classes are included in PSPACE.) Further classes can be obtained by alternating the counting quantifier and the usual existential and universal quantifiers, as in the class $\exists C\forall CP$. Wagner has shown that all these classes have complete sets and, at least for the bottom classes, they look quite natural.

It is well known that the classes of the polynomial-time hierarchy have an equivalent oracle characterization, e.g., $\exists\forall P = NP^{NP}$, $\exists\forall\exists P = NP^{NP^{NP}}$. Torán has shown [To88a,To88b] that the classes of the counting hierarchy have a similar natural oracle characterization, for example,

$$\exists C\forall CP = NP^{PP^{co-NP^{PP}}} = NP^{PP^{NP^{PP}}}.$$

Another similar quantifier $C_=$ can be defined by asking for *exactly* $t(x)$-many strings y in the above definition [Wa86b,To88b]. Although $\forall P \subseteq C_=P$ is easiliy seen, it is not clear whether NP $(= \exists P)$ is included in $C_=P$. Later, it will be shown that a certain subclass of NP, called FewP, is included in $C_=P$. On the other hand, we have the following inclusion.

Proposition 8.2.2 *The class* $C_=P$ *is included in* CP.

Proof. Let $A \in C_=P$. Using dummy computation paths, the threshold for the class $C_=P$ can be assumed to be exactly $t(x) = |\{y \mid |y| \le p(|x|)\}|/2$. That is, $x \in A$ if and only if for exactly half of the potential y's, $(x,y) \in B$, where B is a set in P. It follows, $x \in A$ if and only if for *at least* $t(x)^2$ many pairs (y,z), $(x,y) \in B$ and $(x,z) \notin B$. The language $\{(x,(y,z)) \mid (x,y) \in B \text{ and } (x,z) \notin B\}$ is in P, therefore, A is in CP. □

Whether the inverse inclusion holds is unknown. Torán [To88a] has shown the following inclusions: $\exists CP \subseteq \exists C_=P$ and $C_=CP \subseteq C_=C_=P$.

Let UP $\subseteq$ NP be the class of languages recognizable by polynomial-time nondeterministic Turing machines which have at most one accepting computation path [Va76]. Let US [He87a] be the class of languages L for which there exists a polynomial-time nondeterministic Turing machine M

such that a string x is in L if and only if M on input x has exactly one accepting computation path. Obvious relationships are $UP \cup$ co–$NP \subseteq US$, and $US \subseteq C_=P$.

Discussion. Using several polynomial-length bounded quantifiers, several hierarchies of classes, extending the well-studied polynomial-time hierarchy can be obtained. Some of these classes might turn out to be useful for other structural properties of complexity classes. Results in this direction are presented in Section 4.

Several potential inclusion relationships between such classes involving counting quantifiers remain unresolved, for example, is CP (or at least $\exists P$) included in $C_=P$?

8.3   What To Count

Given the computation tree of a nondeterministic polynomial-time machine, there are subtle differences in what to count. At first, one can count the number of accepting computations of a machine M on input x; let $acc_M(x)$ be this number. Valiant [Va79a,Va79b] introduced the following class of functions

$$\#P = \{f : \Sigma^* \to \mathbb{N} \mid f = acc_M \text{ for some NP machine } M\}$$

Valiant showed that for many NP-complete problems $A = L(M)$, the corresponding counting function acc_M is $\#P$-complete. But, interestingly, this is also true for certain counting functions whose associated decision problem is in P.

The function class $\#P$ is closely related with the language class CP ($=PP$) because $P^{\#P} = P^{CP}$ (a proof can be found in [BBS86]). It is not known whether $\#P$ (resp. CP) is included in the polynomial-time hierarchy. On the other hand, the decision problem that corresponds to a function in $\#P$, is always in NP.

Therefore, for those problems with a $\#P$-complete counting function, counting does not seem to be Turing-reducible to the decision problem.

It has been noticed by Mathon [Ma79] that for the *graph isomorphism* problem counting and deciding are Turing-reducible to each other. This is already some kind of evidence that *graph isomorphism* is not NP-complete. Further evidence for this is given in [Sc87,BHZ87].

In Section 1, we already considered the *census* of a language, another concept of counting. In [BGS87] it is shown that for all unambiguous context-free languages A, $cens_A$ can be computed efficiently. On the other hand, if for some particular ambiguous context-free language B, $cens_B$ could be computed efficiently, then $E = NE$, which implies by [HIS83] that there is no sparse set in $NP - P$.

Related to the census function is the *ranking* function of a language. This subject is addressed in the paper by Li and Vitányi [LV90] contained in this volume. Goldberg and Sipser [GS85] introduced the notion of ranking in the context of efficiently compressing languages. For a set A let $\mathrm{rank}_A : \Sigma^* \to \mathbb{N}$ be the function which, on argument x, gives the number of strings in A which precede x according to lexicographic order, i.e.

$$\mathrm{rank}_A(x) = |\{y \in A \mid y \leq x\}|$$

Define

$$\mathrm{rank\text{-}P} = \{f : \Sigma \to \mathbb{N} \mid f = \mathrm{rank}_A \text{ for some } A \in \mathrm{NP}\}.$$

A further variant of counting is introduced in [KST88a]. Consider nondeterministic machines with accepting and rejecting final states, and additionally equipped with an output tape. The content of the output tape is only considered to be "valid" if the machine stops in an accepting final state. For such a machine M, let $\mathrm{span}_M(x)$ be the number of different valid outputs that M on input x can produce on its various nondeterministic computation paths. Define

$$\mathrm{span\text{-}P} = \{f : \Sigma^* \to \mathbb{N} \mid f = \mathrm{span}_M \text{ for some NP machine } M\}.$$

Theorem 8.3.1 ([KST88a])   (a) *The function class* span-P *has complete members.*

(b) span-P *is included in* #NP *(cf. [Va79a, Va79b]).*

(c) *The function classes* opt-P *[Kr86],* #P, *and* rank-P *are included in* span-P.

The class opt-P was introduced by Krentel [Kr86] in the context of combinatorial optimization problems, and is included in the polynomial-time hierarchy, in level Δ_2^P. The class span-P, like #P, is not known to be in the polynomial hierarchy. Therefore, opt-P and span-P are probably different classes. On the other hand, it turns out that opt-P$[O(\log n)]$ = span-P$[O(\log n)]$ (see [Kr86] and [KST88a] for the notations).

The motivation for introducing the class span-P came from the observation that a function related with the *graph isomorphism* problem is a member of span-P, and not known to be in #P.

Proposition 8.3.2 ([KST88a]) *The function f given by*

$$\bullet \, f(G_1, G_2) = \begin{cases} 2 \cdot n! & \text{if } G_1 \not\cong G_2 \\ n! & \text{if } G_1 \cong G_2 \end{cases}$$

is a member of span-P *(letting n be the number of nodes in the graphs).*

Proof. Consider the following nondeterministic machine M that in input (G_1, G_2), guesses two permutations π_1, π_2 on the set $\{1, \dots, n\}$. Further, M nondeterministically guesses $i \in \{1, 2\}$. Then M verifies that π_1 is an automorphism of the graph $\pi_2(G_i)$. If so, M halts in an accepting state with the output $(\pi_2(G_i), \pi_1)$. It can be seen that $f = \mathrm{span}_M$. $\square$

Although #P and span-P seem closely related, showing equality would give a solution to another important problem in structural complexity theory.

Theorem 8.3.3 ([KST88a])

$$\#P = \text{span-P} \ \textit{if and only if} \ UP = NP.$$

Proof. Suppose $\#P = \text{span-P}$ and let $L = L(M)$ be a set in NP. Define a machine M' that outputs 1 in an accepting final state if and only if M accepts. Then, $\mathrm{span}_{M'}$ is the characteristic function of L which, by assumption, is in #P. This means that $L \in UP$.

Conversely, suppose $UP = NP$ and let f be in span-P, i.e. $f = \mathrm{span}_M$ for a suitable machine M. Then the set $\{(x, y) \mid y \text{ is a valid output of } M \text{ on } x\}$ is in NP, and by the assumption, in UP. Therefore, there is a machine M' which accepts this set, and $acc_{M'}$ takes only values from $\{0, 1\}$. Define another machine M'' which on input x guesses y and simulates M' on input (x, y). Then $f = acc_{M''}$ which proves that f is in #P. $\square$

A consequence of this theorem is that if there is a ranking function for some NP set (i.e. a function in rank-P) that is not in #P then $UP \neq NP$. Notice that $P \neq UP$ if and only if one-way functions exist [GS84].

Another similar equivalence from [KST88a] is that span-P = #NP if and only if NP = co–NP.

Discussion. It could be seen that there are different ways of assigning a counting function to a nondeterministic computation, leading to classes such as #P, #NP, span-P, and rank-P. The question of equality or difference of these function classes can be connected with other open problems in complexity theory, such as UP =? NP.

The class span-P has complete members (under the metric reducibility introduced by Krentel [Kr86]). On the other hand, the status of the function f from Proposition 8.3.2 is still open. Another interesting function is

$$h(G) = \text{ number of Hamiltonian subgraphs of } G.$$

The function h is known to be hard for #P, and located in the class span-P. It is open whether h is span-P-complete.

8.4   Parity

The class $\oplus P$ ("parity-P") was introduced in [PZ83] as a "more moderate version of the counting idea". More generally, for a class of sets $\mathcal{K}$, let us define $\oplus \mathcal{K}$ as the class of set A for which there is a set $B \in \mathcal{K}$ and a polynomial p such that

$$A = \{x \mid \text{the number of strings } y \text{ with } |y| \le p(|x|) \text{ and } (x,y) \in B \text{ is odd}\}.$$

The class $\oplus P$ has been called EP in [GP86], and in this paper it was also observed that the *graph isomorphism* problem restricted to tournament graphs is a member of the class $\oplus P$. The reason is that the automorphism group of a tournament graph is always of odd order. Therefore, if two tournament graphs are non-isomorphic, then there are zero isomorphisms, an even number. On the other hand, if two tournament graphs are isomorphic then the number of isomorphism between them corresponds to the number of automorphisms, thus it is an odd number.

The class $\oplus P$ has obviously complete sets, such as $parity\text{-}SAT = \{F \mid F \text{ is a Boolean formula with an odd number of satisfying assignments}\}$. Unlike the class CP, the class $\oplus P$ is closed under all Boolean operations, and furthermore, $\oplus P = \oplus \oplus P = \oplus P^{\oplus P}$ [PZ83]. Therefore, $\oplus P$ is also closed under polynomial-time Turing reducibility.

The relationship between $\oplus P$ and NP is unclear. Probably both classes are incomparable. Clearly $UP \subseteq \oplus P$. Moreover, Cai and Hemachandra [CH89] have recently shown the following unexpected inclusion relationship (They actually prove a stronger result.) Let FewP be the class of sets A which are accepted by nondeterministic polynomial-time Turing machines which in case of acceptance have a polynomially bounded number of accepting computation paths. Allender [Al85] introduced this class as a natural generalization of UP.

Theorem 8.4.1 ([CH89])

$$\text{FewP} \subseteq \oplus P.$$

Proof. Let M be a machine witnessing that the set A is in FewP. Then $A = L(M)$ and $acc_M(x) \le p(|x|)$ for some polynomial p and all x. Define another machine M' as follows.

> input x ;
> guess a number k, $1 \le k \le p(|x|)$;
> guess a sequence $y_1, \ldots, y_k$ of computation paths for M on x
>   such that $y_1 < \ldots < y_k$;
> If M on x with computation path y_i is accepting for all $i = 1, \ldots, k$ then halt accepting else halt rejecting.

Clearly M' has zero accepting computations if x is not in A, and if x is in A then suppose M has m, $1 \leq m \leq p(|x|)$, many accepting computations. In this case, M' has exactly

$$\sum_{i=1}^{m} \binom{m}{k} = 2^m - 1$$

accepting computations which is an odd number. Hence A is in $\oplus$P.   □

The proof method of the previous theorem has been extended in [KST88b] to show the following inclusion.

Theorem 8.4.2 ([KST88b])

$$\text{FewP} \subseteq \text{C}_{=}\text{P}.$$

Proof. Modify the last statement of the above algorithm for M' as follows.

If [(M on x with computation path y_i is accepting for all i) iff (k is odd)] then halt accepting else halt rejecting.

The analysis of this modified M' shows that $M'(x)$ has either $g(n)$ or $g(n) + 1$, $n = |x|$, accepting computations, for some polynomial-time computable (but not polynomially bounded) function g. The former happens in case $x \notin A$, the latter in case $x \in A$.   □

A consequence of Theorem 8.4.2 is that the class FewP is "low" for the operators $\oplus$P, CP, and C$_=$P, that is, $\oplus$P^{FewP} = $\oplus$P, CP^{FewP} = CP, and C$_=$P^{FewP} = C$_=$P.

A recent important result by Toda [Tod89] shows that $\oplus$P is ($\leq_T^{\text{BPP}}$-) hard for the polynomial-time hierarchy, and furthermore,

$$\text{PP}^{\text{PH}} \subseteq \text{P}^{\text{PP}}.$$

Further results, comparing $\oplus$P with other structural notions can be found in [He87d,GJY87].

Discussion. The parity operation leads to further classes loosely connected with counting. The operator $\oplus$ can be mixed with the operators $\exists$, $\forall$, C leading to further classes that have not yet been analyzed thoroughly. As an example, it is known that the class BPP [Gi77] is included in $\exists\forall$P $\cap$ $\forall\exists$P [La83,Si83]. The known proofs of this fact use parity operations (i.e. XORing) as a technical tool. This is an indication that possibly an inclusion relation such as BPP $\subseteq$ $\oplus\exists$P or BPP $\subseteq$ $\exists\oplus$P might hold. But this is still open.

Further questions worth analyzing include classes defined in terms of parity-rank, parity-span, parity-opt, etc.

The class $\oplus P$ is closed under complementation, and probably incomparable with the class NP. This suggests consideration of the following questions: Is NP $\cap$ co–NP included in $\oplus P$? Is NP $\cap$ co–NP $\cap$ $\oplus P$ equal to P (cf. [GP86])? Is NP $\cap$ $\oplus P$ a member of the *low hierarchy* (see [Sc83,Sc87])?

8.5   Approximate Counting

If a function f cannot be feasibly computed, then one might try to approximate f using feasible resources. More precisely, even if f is not computable in polynomial time, there might exist a function g – computable with fewer resources than needed for f – such that

$$(1 - e) \cdot g(x) \leq f(x) \leq (1 + e) \cdot g(x)$$

where e is a small constant, or even a function depending on x that approaches 0 as $|x|$ grows (for example, $e = O(|x|^{-t})$ for some constant t). Notice that the size of the set

$$\{n \in \mathbb{N} \mid (1 - e) \cdot g(x) \leq n \leq (1 + e) \cdot g(x)\}$$

still can grow exponentially in $|x|$.

Stockmeyer [St85] has shown that every function $f \in \#P$ can be approximated in the above sense (with $e = O(|x|^{-t})$) where the computation of the function g needs only resources of the type $\Delta_3^P = P^{\exists \forall P}$ (actually only $P^{\exists \forall P}[\log]$ where [log] indicates a logarithmic bound on the number of oracle queries, see [Pi88]). His proof uses Sipser's Coding (or Hashing) Lemma from [Si83].

Theorem 8.5.1 ([St85]) *For every function $f \in \#P$ and every polynomial q there is a function $g \in P^{\exists \forall P}[\log]$ such that for all x, $|x| = n$,*

$$(1 - 1/q(n)) \cdot g(x) \leq f(x) \leq (1 + 1/q(n)) \cdot g(x).$$

Proof (sketch). Let $f \in \#P$, i.e. $f = acc_M$ for some NP machine M. Let p be the polynomial time bound of M. Let $ACC_M(x) \subseteq \{0,1\}^{p(|x|)}$ be the set of accepting computations of M on input x (described as 0-1-strings), then $acc_M(x) = |ACC_M(x)|$. Consider the following oracle set

$$A = \{(x, m) \mid \text{there is a set of linear transformations } h_1, \ldots, h_m$$
$$\text{from } \{0, 1\}^{p(|x|)} \text{ to } \{0, 1\}^m \text{ (given as a Boolean } p(|x|) \times m$$
$$\text{matrices) which "isolate" (see [Si83]) every } y \in ACC_M(x) \}.$$

This set A is in $\exists \forall P$ since it can be described in the form

$$A = \{(x,m) \mid \exists h_1,\ldots,h_m \; \forall y \; \exists i \; \forall z \; [(y \in ACC_M(x) \wedge z \in$$
$$ACC_M(x) \wedge y \neq z) \rightarrow h_i(y) \neq h_i(z)] \}$$

Notice that the second existential quantifier has only polynomial range and thus can be eliminated.

It is clear that if $acc_M(x) = |ACC_M(x)| > m \cdot 2^m$ then there must exist a $y \in ACC_M(x)$ that are not isolated, therefore $(x,m) \notin A$. On the other hand, if $acc_M(x) = |ACC_M(x)| < 2^{m-1}$, then it can be shown that such linear mappings exist, i.e. $(x,m) \in A$.

Therefore, a polynomial-time algorithm using A as oracle set can proceed as follows. On input x, using binary search, the algorithm finds the minimum m such that $(x,m) \in A$ (therefore, $(x,m-1) \notin A$). This can be done by logarithmically (in $|x|$) many queries to the oracle. Then, by the above estimations, the value of $f(x)$ is in the interval $[2^{m-2}, m \cdot 2^m]$. This is a rough estimation for $f(x)$ which does not meet the requirement in the statement of the Theorem. Now the following trick works: Instead of estimating f directly, as described above, we estimate the function $h(x) = f(x)^{r(|x|)}$ for some an appropriate polynomial r (depending on q). This function is still in #P. From a similar rough estimation to h, a sharper estimation for f can be obtained which meets the requirements of the Theorem. $\square$

Jerrum, Valiant, and Vazirani [JVV86] (see also [SJ87]) found that if the basis oracle algorithm is allowed to behave probabilistically, then an NP (or co–NP) oracle suffices. In this case, the algorithm itself guesses probabilistically the linear transformations $h_1,\ldots,h_m : \{0,1\}^{p(|x|)} \rightarrow \{0,1\}^m$, for various values of m, and then queries the following oracle $B \in$ co–NP.

$$B = \{(x,h_1,\ldots,h_m) \mid h_1,\ldots,h_m \text{ are linear transformations}$$
from $\{0,1\}^{p(|x|)}$ to $\{0,1\}^m$, $m \geq 0$, given as a Boolean
$p(|x|) \times m$ matrices, which isolate every $y \in ACC_M(x) \}$

The price to be paid is that the outputted approximation value for $f(x)$ is in the desired interval just with high probability, but not for certain. More formally stated, the result is the following.

Theorem 8.5.2 ([JVV86]) *For every function $f \in$ #P and all polynomials p, q there is a function $g \in \text{BPP}^{\text{NP}}[\log]$ such that for all x, the inequality*

$$(1 - 1/q(n)) \cdot g(x) \leq f(x) \leq (1 + 1/q(n)) \cdot g(x)$$

holds with probability at least $1 - 2^{-p(|x|)}$.

In [VV86] it is shown that the above approximation algorithms (for Theorem 8.5.1 and 8.5.2) can be somewhat simplified – but for the price of a more complicated analysis.

In [KST88a] it is shown that Theorems 8.5.1 and 8.5.2 can be extended to the class span-P instead of #P. Therefore, these approximations are applicable to the function f presented in Proposition 8.3.2. Such approximations play a crucial role in the recent result that the *graph isomorphism* problem is in the low hierarchy [Sc87].

A different type of approximation for #P functions is considered in [CH88]. There, an approximation algorithm for a function $f \in$ #P is understood as a deterministic procedure which outputs a polynomial size set of values among which the correct value $f(x)$ is guaranteed to occur. It is shown that this type of approximation is as hard as computing f exactly.

Discussion. We have seen that functions of counting classes can be approximated by functions in the polynomial hierarchy. This is remarkable because the class #P is not known to be included in the polynomial hierarchy. An interesting question remains: What is the relationship between these types of approximations discussed here, and the approximation schemes for NP-complete optimization problems.

8.6   Relativizations

Relativization results are usually taken as a tool to indicate the difficulty of a certain problem in the unrelativized case. The most prominent example is the "double relativization" of the P=NP problem, due to Baker, Gill, and Solovay [BGS75]: There are oracles A and B such that $P^A = NP^A$, but $P^B \neq NP^B$. Therefore, non-relativizable techniques are needed to solve the P=NP problem.

Concerning counting classes, the first reference for a relativization result is [An80] showing that there is an oracle A under which PP ($=$ CP) is not included in the classes $\exists \forall P$ and $\forall \exists P$ of the polynomial-time hierarchy.

Yao [Ya85] and Hastad [Ha87] have shown that there are oracles which separate PSPACE from the polynomial-time hierarchy. But actually, the "test language" that is used in these proofs is

$$L(A) = \{0^n \mid \text{there is an odd number of strings of size } n \text{ in } A\}$$

This language is obviously in $\oplus P^A$, hence the oracle A separates not only PSPACE from the polynomial hierarchy PH, but also $\oplus P$ from PH. Actually, the proof shows that the parity function needs at least exponential-size to be computed by constant-depth depth circuits. The relativization result then falls out using an observation due to Furst, Saxe, and Sipser [FSS84].

Similar bounds as for the parity function can be shown for the majority function [Ha87]. Therefore, it follows that there is an oracle separating CP from PH.

Recently, some new separation results between counting classes have been shown by Torán [To88a].

Theorem 8.6.1 *There exist oracles for each of the following situations.*

    (a) NP *is not included in* ⊕P.

    (b) NP *is not included in* $C_=P$.

    (c) CP *is not included in* $C_=P$.

    (d) $C_=P$ *is not closed under complementation.*

    (e) CP *and* ⊕P *are incomparable.*

    (f) CP *is different from* ∃CP *and from* ∀CP.

Proof (sketch) for (a). We proceed as in Baker, Gill, Solovay's proof [BGS75] separating NP from P. For any set A let

$$L(A) = \{0^n \mid \text{there is a string of size } n \text{ in } A\}$$

Then, clearly, $L(A)$ is in NP^A for any A. Now, a particular oracle A is defined by a stage-by-stage construction so that $L(A)$ is not in $\oplus P^A$.

Let $M_1, M_2, \ldots$ be an enumeration of the nondeterministic polynomial-time oracle machines with corresponding polynomial running times $p_1, p_2, \ldots$.

    Stage 0:
    Let $A_0 = \emptyset$ and $k_0 = 0$.

    Stage $n > 0$:
    Let $k_n > k_{n-1}$ be the smallest integer m satisfying

$$m > p_{n-1}(k_{n-1}) \text{ and } 2^m > p_n(m);$$

    (*) Choose a set $B \subseteq \Sigma^{k_n}$ such that

$$B = \emptyset \iff M_n \text{ on input } 0^{k_n} \text{ with oracle } A_{n-1} \cup B \text{ has}$$
$$\text{an odd number of accepting computations ;}$$

Finally define A as the union of all A_n.

    Obviously, the crucial point in this construction is the choice of the the set B in line (*). The existence of such B can be guaranteed by a tricky counting argument that can be found in [To88a].   □

    This separation result should be contrasted with a recent "positive relativization result" (cf. [LS86,BBS86]) with respect to sparse oracles that is shown in [KST88b].

Theorem 8.6.2   *(a)* NP $= \oplus$P *if and only if for every sparse oracle set S,* $NP^S = \oplus P^S$.

(b) NP $=$ C$_=$P *if and only if for every sparse oracle set S*, NP^S
$=$ C$_=$P^S.

(c) $\oplus$P $=$ CP *if and only if for every sparse oracle set S*, $\oplus$P^S
$=$ CP^S.

Focusing on the classes NP and $\oplus$P, the theorem says that these classes cannot be separated with a sparse oracle unless the classes are different in the unrelativized case. Indeed, the counting argument mentioned in the proof of Theorem 8.6.1 leads to a non-sparse oracle that separates $\oplus$P from NP.

8.7 REFERENCES

[Al85]   E. Allender. *Invertible Functions.* Ph.D. thesis, Georgia Tech., 1985.

[An80]   D. Angluin. On counting problems and the polynomial-time hierarchy. *Theor. Comput. Sci.* 12 (1980): 161–173.

[BGS75]   T.P. Baker, J. Gill, and R.M. Solovay. Relativizations of the P=?NP question. *SIAM J. Comput.* 4 (1975): 431–442.

[BBS86]   J.L. Balcázar, R.V. Book, and U. Schöning. The polynomial-time hierarchy and sparse oracles. *Joun. of the Assoc. Comput. Mach.* 33 (1986): 603–617.

[Ba68]   Y.M. Barzdin. Complexity of programs to determine whether natural numbers not greater than n belong to a recursively enumerable set. *Soviet Math. Dokl.* 9 (1968): 1251–1254.

[BGS87]   A. Bertoni, M. Goldwurm, and M. Sabatini. Computing the counting function of context-free languages. *Symp. Theor. Aspects Computer Science*, Lecture Notes in Computer Science 247, 169–179, Springer-Verlag, 1987.

[BHZ87]   R.B. Boppana, J. Hastad, and S. Zachos. Does co-NP have short interactive proofs? *Inform. Proc. Letters* 25 (1987): 27–32.

[CH88]   J. Cai and L.A. Hemachandra. Enumerative counting is hard. *Proc. 3rd Structure in Complexity Theory Conf.*, 194–203, IEEE, 1988.

[CH89]   J. Cai and L.A. Hemachandra. On the power of parity. *Symp. Theor. Aspects of Comput. Sci.*, Lecture Notes in Computer Science, Springer-Verlag, 1989, to appear.

[FSS84] M. Furst, J.B. Saxe, and M. Sipser. Parity, circuits, and the polynomial-time hierarchy. *Math. Syst. Theory* 17 (1984): 13–27.

[Gi77]   J. Gill. Computational complexity of probabilistic complexity classes. *SIAM Journ. Comput.* 6 (1977): 675–695.

[GS85]   A.V. Goldberg and M. Sipser. Compression and ranking. *17th ACM Symp. Theory Comput.* 440–448, 1985.

[GP86]   L.M. Goldschlager and I. Parberry. On the construction of parallel computers from various bases of boolean functions. *Theor. Comput. Sci.* 43 (1986): 43–58.

[GJY87] J. Goldsmith, D. Joseph, and P. Young. Using self-reducibility to characterize polynomial time. *Tech. Report* 87-11-11, Computer Science Dept., Univ. of Washington, Seattle, 1987.

[GS84]   S. Grollmann and A.L. Selman. Complexity measures for public-key crypto-systems. *25th Symp. Found. Comput. Sci.*, 495–503, IEEE, 1984.

[HIS83] J. Hartmanis, N. Immerman and V. Sewelson. Sparse sets in NP-P: EXPTIME versus NEXPTIME. *15th ACM Symp. Theory Comput.*, 382–391, 1983.

[Ha87]   J.T. Hastad. *Computational limitations for small-depth circuits.* Ph.D. thesis, MIT Press, Cambridge, MA., 1987.

[He87a] L.A. Hemachandra. *Counting in Structural Complexity Theory.* Ph.D. thesis, Cornell University, 1987.

[He87b] L.A. Hemachandra. On ranking. *Proc. 2nd Structure in Complexity Theory Conf.*, 103–117, IEEE, 1987.

[He87c] L.A. Hemachandra. The strong exponential hierarchy collapses. *19th ACM Symp. Theory Comput.*, 110–122, 1987.

[He87d] L. A. Hemachandra. *On parity and near-testability: $P^A \neq NT^A$ with probability 1.* Tech. Report 87-11-11, Comput. Sci. Dept., Univ. of Washington, Seattle, 1987.

[Hu88]   D.T. Huynh. The complexity of ranking. *Proc. 3rd Structure in Complexity Theory Conf.*, 204–212, IEEE, 1988.

[Im88]   N. Immerman. Nondeterministic space is closed under complement. *Proc. 3rd Struct. Complexity Theory Conf.*, 112–115, IEEE, 1988.

[JVV86]   M.R. Jerrum, L.G. Valiant, and V.V. Vazirani. Random generation of combinatorial structures from a uniform distribution. *Theor. Comput. Sci.* 43 (1986): 169–188.

[Ka87]   J. Kadin. $P^{NP[\log n]}$ and sparse Turing complete sets for NP. *Proc. 2nd Struc. in Complexity Theory Conf.*, 33–40, IEEE, 1987.

[KL80]   R.M. Karp and R.J. Lipton. Some connections between nonuniform and uniform complexity classes. *Proc. 12th ACM Symp. Theory of Comput. Sci.*, 302–309, 1980.

[Ko88]   K. Ko. Relativized polynomial time hierarchies having exactly k levels. *Proc. 3rd Structure in Complexity Theory*, 251, IEEE, 1988.

[KST88a] J. Köbler, U. Schöning, and J. Torán. On counting and approximation. *Proc. Colloq. Trees in Algebra and Programming* 1988, Lecture Notes in Computer Science 299, 40–51, Springer-Verlag, 1988.

[KST88b] J. Köbler, U. Schöning, and J. Torán. Turing machines with few accepting paths. *manuscript*, 1988.

[Kr86]   M.W. Krentel. The complexity of optimization problems. *18th ACM Symp. Theory Comput.*, 69–76, 1986.

[La83]   C. Lautemann. BPP and the polynomial hierarchy. *Inform. Proc. Letters* 14 (1983): 215–217.

[Lo82]   T.J. Long. Strong nondeterministic polynomial-time reducibilities. *Theor. Comput. Sci.* 21 (1982): 1–25.

[LS86]   T.J. Long and A.L. Selman. Relativizing complexity classes with sparse sets. *Journ. of the Assoc. Comput. Mach.* 33 (1986): 618–628.

[LV90]   M. Li and P. Vitányi. Applications of Kolmogorov Complexity in the Theory of Computation. In A. Selman, editor *Complexity Theory Retrospective*, pages 147–203, Springer-Verlag, 1990.

[Ma82]   S.A. Mahaney. Sparse complete sets for NP: solution of a conjecture of Berman and Hartmanis. *Journ. Comput. Syst. Sci.* 25 (1982): 130–143.

[MH80]   S.A. Mahaney and J. Hartmanis. An essay about research on sparse NP complete sets. *Math. Found. Computer Science* 1980,

Lecture Notes in Computer Science 88, 40–57, Springer-Verlag, 1980.

[Ma79]   R. Mathon. A note on the graph isomorphism counting problem. *Inform. Proc. Lett.* 8 (1979): 131–132.

[PZ83]   C.H. Papadimitriou and S.K. Zachos. Two remarks on the power of counting. *6th GI Conf. on Theor. Comput. Sci.*, Lecture Notes in Computer Science 145, 269–276, Springer-Verlag, 1983.

[Pi88]   M. Piotrów. On the complexity of counting. *Symp. Math. Found. Comput. Sci.*, Lecture Notes in Compuyter Science 324, 472–482, Springer-Verlag, 1988.

[Sc83]   U. Schöning. A low and a high hierarchy within NP. *Journ. Comput. Syst. Sci.* 27 (1983): 14–28.

[Sc87]   U. Schöning. Graph isomorphism is in the low hierarchy. *4th Symp. Theor. Aspects of Comput. Sci.*, Lecture Notes in Computer Science 247, 114–124, Springer-Verlag, 1987.

[SW88]   U. Schöning and K.W. Wagner. Collapsing oracle hierarchies, census functions, and logarithmically many queries. *Symp. Theor. Aspects Computer Science* 1988, Lecture Notes in Computer Science 294, 91–97, Springer-Verlag, 1988.

[Si77]   J. Simon. On the difference between one and many. *Intern. Conf. Automata, Lang., Progr.* 1977, Lecture Notes in Computer Science 52, 480–491, Springer-Verlag 1977.

[SJ87]   A. Sinclair and M. Jerrum. Approximate counting, uniform generation and rapidly mixing Markov chains. *Internal Report* CSR-241-87, Department of Computer Science, University of Edinburgh, 1987.

[Si83]   M. Sipser. A complexity theoretic approach to randomness. *Proc. 15th ACM Symp. Theory of Comput. Sci. 1983*, 330–335.

[St77]   L. J. Stockmeyer. The polynomial-time hierarchy. *Theor. Comput. Sci.* 3 (1977): 1–22.

[St85]   L.J. Stockmeyer. On approximation algorithms for #P. *SIAM Journ. Comput.* 14 (1985): 849–861.

[Sz87]   R. Szelepcsényi. The method of forcing for nondeterministic automata. *Bulletin EATCS* 33 (1987): 96–99.

[Tod87]   S. Toda. Σ_2SPACE(n) is closed under complement. *Journ. Comput. Syst. Sci.* 35 (1987): 145–152.

[Tod89]   On the computational power of PP and $\oplus$P. *30th Symp. Found. Comput. Sci.*, 514–519, IEEE, 1989.

[To88a]   J. Torán. *Structural Properties of the Counting Hierarchies.* Doctoral dissertation, Facultat d'Informatica, UPC Barcelona, Jan. 1988.

[To88b]   J. Torán. An oracle characterization of the counting hierarchy. *Proc. 3rd Struct. Complexity Theory Conf.*, 213–223, IEEE, 1988.

[Va76]   L.G. Valiant. The relative complexity of checking and evaluating. *Inform. Proc. Lett.* 5 (1976): 20–23.

[Va79a]   L.G. Valiant. The complexity of computing the permanent. *Theor. Comput. Sci.* 8 (1979): 181–201.

[Va79b]   L.G. Valiant. The complexity of reliability and enumerability problems. *SIAM Journ. Computing* 8 (1979): 410–421.

[VV86]   L.G. Valiant and V.V. Vazirani. NP is as easy as detecting unique solutions. *Theor. Comput. Sci.* 47 (1986): 85–93.

[Wa86a]   K.W. Wagner. Some observations on the connection between counting and recursion. *Theor. Comput. Sci.* 47 (1986): 131–147.

[Wa86b]   K.W. Wagner. The complexity of combinatorial problems with succinct input representation. *Acta Inform.* 23 (1986): 325–356.

[Wr77]   C. Wrathall. Complete sets and the polynomial-time hierarchy. *Theor. Comput. Sci.* 3 (1977): 23–33.

[Ya85]   A. Yao. Separating the polynomial-time hierarchy by oracles. *26th Proc. Found. Comput. Sci.*, 1–10, IEEE, 1985.

Author Index

Subject Index